Never Trust Your Eyes

The world *appears* to us in splendid detail, in glorious color, in three dimensions and full motion. So we might reasonably think that our sense of sight is near-perfect.

Yet the reality is that we only see detail over a tiny area, we are legally blind 25 degrees off center, and we are completely blind for up to half the time our eyes are open!

Using striking visual illusions, we demonstrate these huge limitations, and compare the eye to a modern digital camera that can take both stills and movies.

Along the way, you will learn about the eye and camera, and how Nature ingeniously provides us with the ultimate illusion of near-perfect sight, well matched to our survival in the real world.

Acknowledgements

Thanks to Dr R W G Hunt, Dr C A Padgham, and Dr J E Saunders for their original and inspirational lectures on perception and color.

Thanks to my wife Valerie for her continuing support and encouragement, and to enduring friends and relatives who have acted as target audience, and provided valuable feedback.

Never Trust Your Eyes

Trevor A White   MSc MBA BEng

Never Trust Your Eyes
First Edition

Copyright © 2011 by Trevor A White
All rights reserved

Library of Congress Catalog Number: 2010942219

EAN (ISBN-13): 978-145-64186-6-3

ISBN-10: 145-641866-1

VISAGE BOOKS, Atlanta, GA, USA

Printed in the United States of America

Contents

Introduction

As we look around us, we *perceive* our world in glorious detail, full color, full motion and three dimensions.

So we might reasonably think that our sense of sight is near-perfect.

The reality is that the eye is astonishingly limited in its capabilities, and a continual guessing game goes on between the eye and brain to best interpret what we are seeing at any one moment.

Nevertheless, I want to emphasize that the eye and brain together - the human visual system - form a marvelous instrument, well-adapted to meeting our survival needs in the real, living world.

So how well can we really see?

Let's start by looking at this example of a detailed, colorful scene, captured with a 6 megapixel digital camera.

Figure 1: A digital camera's uniformly detailed, colorful image

Think about what your eyes did as soon as the picture appeared.

Where did they look first?

How did they move around?

How did you build up an impression of the scene?

Figure 2: Illustration of what the eye sees instantaneously

Let's assume you started by looking at the left face. The figure above illustrates what your eye really sees instantaneously.

Notice how only the left face is sharply detailed.

As we move away from this detailed area, the picture becomes more and more blurred.

And we have a real gap in our vision, illustrated here by the gray oval superimposed over the face on the right.

To be precise, this describes what the right eye sees, with its blind spot to the right of the detailed region. The left eye is a mirror image of this, with its blind spot on the left of the detailed region.

Consequently what we see appears amazingly limited compared with the full color, full detail picture we obtained instantly from our digital camera - even though the eye has about six million color photoreceptors, the same total number as the camera we used.

Yet we are completely unaware of these astonishing limitations.

In the following chapters we will look more closely at the eye and explore these limitations - while showing how Nature astutely overcomes them to provide us with a sense of sight well matched to the real world in which we live.

To improve our understanding of how the eye works, we will illustrate the basics of a modern digital camera, that is able to take both stills and movies, and then compare and contrast the man-made camera with the choices made by Nature for our very human sense of vision.

The Basics

To help us better grasp how the eye produces an image of the scene we're viewing, let's first go back in history and briefly look at the how the camera came about.

By the way, did you know the word *camera* comes from Latin, and simply means a *room* or a *chamber*?

The word *obscura*, from *obscurum* meaning *darkness*, was added to create the term *camera obscura* (a dark room or chamber).

The Camera Obscura

Here is a drawing of an early camera obscura. It is simply a light-sealed box, or chamber, with a pinhole to allow light from the scene to enter and create a smaller, inverted image on the inside back surface.

Figure 3: Sketch of an early camera obscura (*Fizyka,1910*)

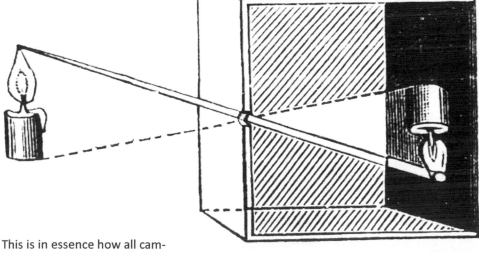

This is in essence how all cameras work - living or engineered.

The camera obscura can literally be a full-size room, and during the 19th century it was quite fashionable to have one as a lookout tower - for example at seaside resorts around Britain's coast.

By placing a mirror diagonally overhead, the scene outside is projected down onto a circular table - see the illustration on the next page.

The audience stands round the table and sees a panoramic view of the beach, town and sea, without the image appearing upside down because the audience simply moves round the table to view the image the right way up.

The mirror was often rotatable, enabling the audience to see all around the town.

So, well before color photography or movie cameras, the camera obscura gave the public their first taste of live, panoramic, full-color, moving images.

Of course it would take many decades of slow and laborious technical advances before we could record and then replay such full color, moving images.

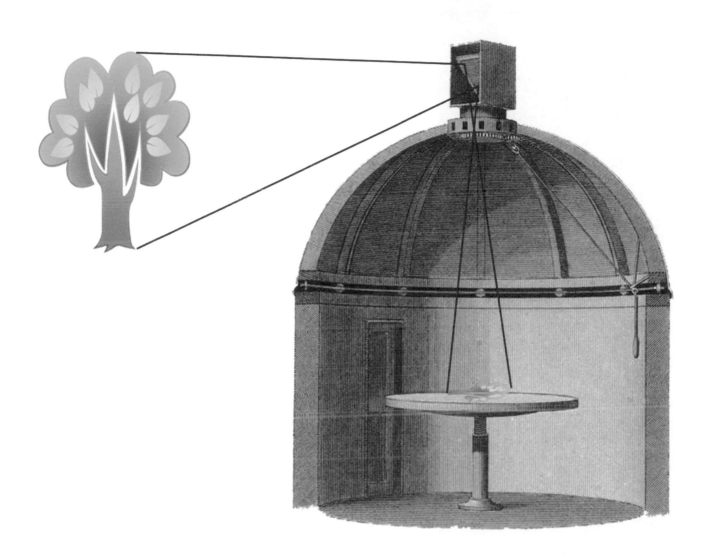

Figure 4: Illustration of a full-size camera obscura with a panoramic rotating mirror *(adapted from A. Rees, Cyclopoedia Universal Dictionary of Arts and Sciences 1778, © National Maritime Museum, Greenwich)*

Figure 5: The wonderfully ornate architecture of the full-size camera obscura in Edinburgh, Scotland *(© Camera Obscura and World of Illusions, Edinburgh)*

Figure 6: The splendid view of the outside from the inside of the Edinburgh camera obscura *(© Camera Obscura and World of Illusions, Edinburgh)*

The Basics of the Eye

The eye operates much like the camera obscura. Light enters the eye through the pupil, which is simply the circular opening provided by the iris, and appears black.

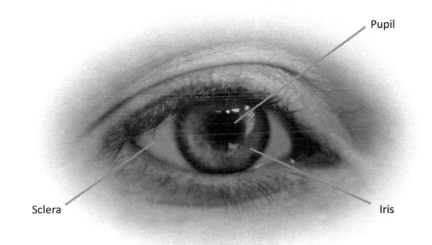

Figure 7: Human eye front view

When we look at a scene, light is focused onto the inside back surface to produce a small, upside-down and laterally-inverted image, just like the camera obscura.

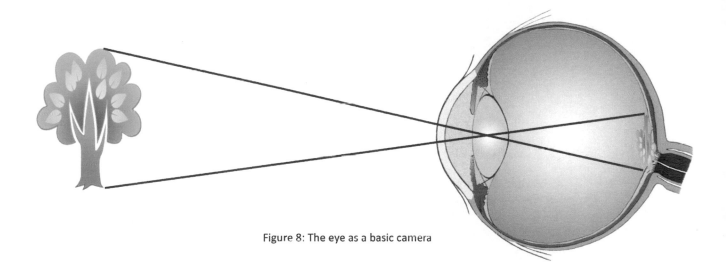

Figure 8: The eye as a basic camera

Looking at a cross-section of the eye, we can see that it comprises these basic elements:

- A mostly light-proof enclosure (Sclera) with a transparent front opening (Cornea)

- An aperture ring around the lens to help control the amount of light entering the eye (Iris)

- A lens to focus light from the scene onto the image sensor

- A thin curved, light-sensitive layer on the inside back surface of the eye that is the image sensor (Retina)

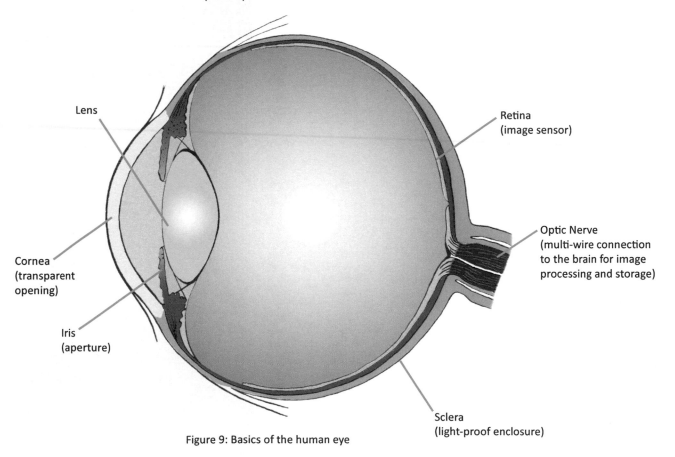

Lens

Retina
(image sensor)

Optic Nerve
(multi-wire connection
to the brain for image
processing and storage)

Cornea
(transparent
opening)

Iris
(aperture)

Sclera
(light-proof enclosure)

Figure 9: Basics of the human eye

The captured image is then sent to the brain via the optic nerve - essentially a multi-wire cable - for further processing and storage.

By the way, when you go to the optometrist, he may put drops in your eyes. The drops contain Atropine and cause the aperture, or iris, to open up as far as possible. This helps the optometrist to better examine the interior of the eye.

Atropine originally came from the plant commonly known as Belladonna.

This means beautiful woman in Italian, and the name comes from its early use by young women to enlarge the pupils of their eyes to make them look more attractive.

It is also commonly known as Deadly Nightshade.

Warning - this plant is poisonous!

Figure 10: Atria Belladonna
(adapted from Koehler's Medicinal-Plants 1887)

(a) Girl with small pupils

(b) Girl with large pupils

Figure 11: The same image but with small
pupils in the upper picture, and large pupils
in the lower picture - is there a difference in
attractiveness?

The Basics of the Camera

A modern-day still or video camera also operates much like the camera obscura.

Figure 12: A modern digital compact camera

When the shutter is open, light enters through the lens and the scene is reproduced as a small, upside-down, laterally-inverted image on the sensor near the inside back surface of the camera.

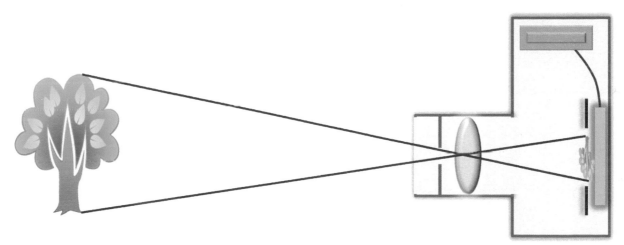

Figure 13: Image creation in a modern-day digital camera

Looking at the cross-section of the camera, we can see that it comprises these basic elements:

- A light-proof enclosure with a transparent front opening

- An aperture ring around the lens to help control the amount of light entering the camera

- A lens to focus light from the scene onto the image sensor

- A light-sensitive area that is the image sensor

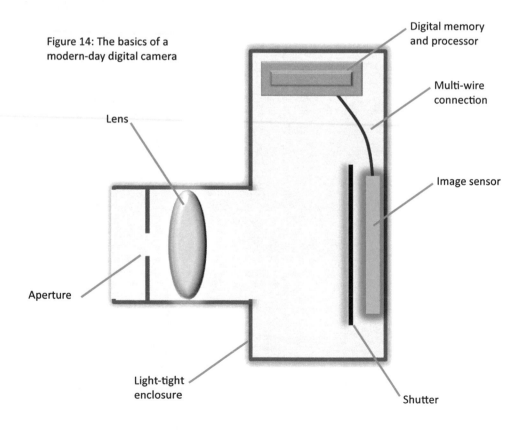

Figure 14: The basics of a modern-day digital camera

Digital memory and processor

Multi-wire connection

Image sensor

Lens

Aperture

Light-tight enclosure

Shutter

So far then we have described elements similar to those of the human eye.

However, a camera has a further required feature:

- A shutter to allow light into the camera only when a picture is to be taken (often referred to as the speed or duration of it being open, e.g. 1/60 second).

In modern cameras, the shutter may be electronic, rather than mechanical.

We shall see later why a camera requires this feature while the human eye does not, and how this feature impacts the pictures a camera produces.

Similar to the eye's optic nerve connecting the retina to the brain, a camera's image sensor is connected to a processor and memory unit.

A video camera operates in essentially the same way as a still camera except that the manual shutter of a still camera is replaced with an automatic shutter that opens and closes many times a second to produce a continuous sequence of still images.

We'll explore video cameras and the eye's perception of movement later.

So How Do Our Eyes Compare With a Camera?

So far, there seems to be very little difference between how the eye works and how a modern-day digital camera works.

Yet, as we pointed out in the Introduction, there are huge differences between the resultant picture we see of the world - our eyes only seeing detail in a very small area of the scene, while the remainder is blurred, except for an oval where we are in fact completely blind.

Thus far, we have only touched on the differences for still images - as a video camera, the eye has a further astonishing limitation.

It is completely blind for up to half the time the eyes are open - and this is not related to blinking.

Again, we are rarely aware of any of these limitations, and have the illusion of near-perfect vision.

In the following chapters we will explore each of these limitations in turn. We will show that what we perceive depends on a myriad of aspects - including where we're concentrating our gaze at any one moment, what exists around the area of interest, what we have seen immediately before, and what prior knowledge and experience we bring to bear.

The Eye's Blind Spots

You don't believe you have a big hole in your vision? Try this simple exercise based on our reference picture.

Figure 15: Illustration for the eye's blind spot

Hold the page at arm's length and close your left eye.

With your right eye, look directly at the left face (the white dot on the nose of the left face is there to help you concentrate).

While concentrating on the left face, slowly bring the page towards you.

At about 30 cm (1 foot) away, you should be able to observe the right face start to disappear.

With practice, and a steady eye, you can easily make the right face completely disappear.

The eye's blind spot has a simple physical explanation. Where the optic nerve leaves the eye on its journey to bring images to the brain, there are no light sensitive receptors and so no vision.

Looking at our cross-section of the eye, we can see where the blind spot is located.

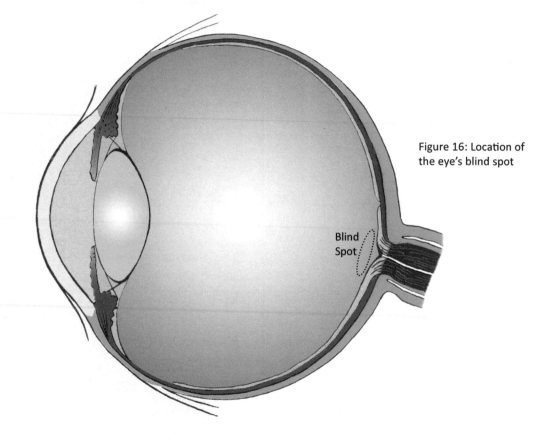

Figure 16: Location of the eye's blind spot

In fact, the blind spot is oval in shape and quite large.

To demonstrate the real size and shape of the blind spot, try these next exercises.

As before, close your left eye and look with your right eye at the center of the star.

Holding the book at arm's length, with your right eye look steadily at the center of the star.

Now bring the book slowly towards you.

With practice, you should be able to make all the colored rings on the right disappear.

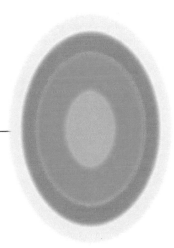

Figure 17: The true size and shape of the eye's blind spot

By the way, if you're more comfortable using your left eye for these exercises, simply close your right eye, and view the figures upside down.

Even more amazing, if we make the surrounding background a pattern, then the brain fills in and extends the pattern into the space where the blind spot occurs, rather than leaving a gap in our vision.

You can try this with the next figure.

Figure 18: Blind Spot - demonstration
of the brain filling in our gap in vision

So why aren't we normally aware of these gaps in our sight?

Looking at the next figure, we can see that the right eye is a mirror image of the left eye. So the blind spots are on opposite sides.

When we look at a scene with both eyes, we have a blind spot that blanks out an oval that is on opposite sides of the scene. You can see this illustrated in the figure where the image from each eye has a grayed out oval showing the blind spot.

Remember from the *camera obscura* that the image at the back of the eye is upside down and reversed left-right.

Now, when we overlay the two images from the eyes, we have all the information needed to fill in the scene

correctly, a little like a modern-day camera can stitch together multiple images to create a single panoramic view.

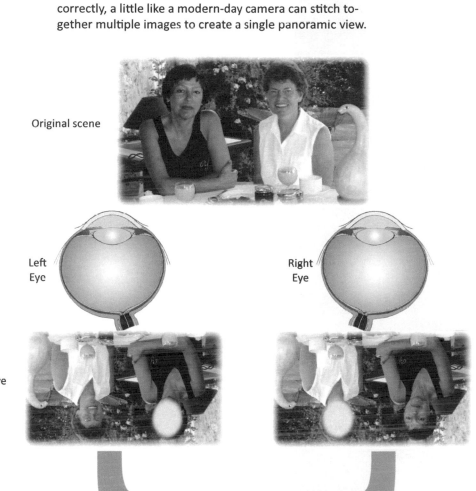

Original scene

Left
Eye

Right
Eye

Image from each eye
with its blind spot

Figure 19: The blind spot in the
right eye is on the opposite side
from that of the left eye

Reconstructed image

Of course, if it were as simple as this, you would see a blind spot with only one eye open, which of course you don't. So something else must be going on.

In fact, each eye is in continual motion, and so the blind spot is being continually filled in. We will explain this motion later, when we cover blur and the camera's need for a shutter.

You might wonder why our eyes have a blind spot at all.

If we look closely at a tiny section of the eye's image sensor, we can see that the nerve fibers - the wiring - from the individual receptors to the optic nerve, along with the blood vessels and preprocessing cells, are IN FRONT OF the eye's image sensor. These all come together in a bundle to exit the eye through the optic nerve.

The optic nerve exits *within* the overall light sensitive surface of the eye, and not to one side. So where this very dense collection of blood vessels and nerve fibers exits, there are no light sensitive receptors, and so we have a physical hole in our light sensitive surface, which we call the blind spot.

In fact, light from the scene has to penetrate the blood vessels, optic nerve fibers, and pre-processing neurons except in one extremely small area called the fovea that we shall describe shortly.

We are completely unaware of having to see through this spider's web of nerves and blood vessels - just as we are completely unaware of having a hole in our vision.

A camera doesn't have these challenges because the wiring to the processor and memory is in the background, not obscuring the light from reaching the image sensor's receptors directly, and doesn't exit as a bundle within the sensitive area of the sensor.

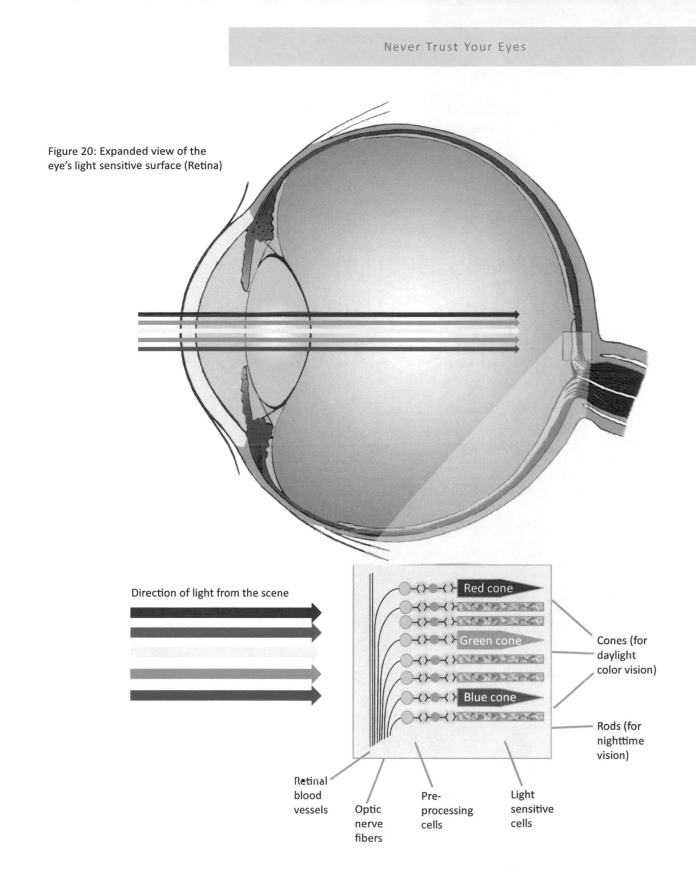

Figure 20: Expanded view of the eye's light sensitive surface (Retina)

Direction of light from the scene

Red cone

Green cone

Blue cone

Cones (for daylight color vision)

Rods (for nighttime vision)

Retinal blood vessels

Optic nerve fibers

Pre-processing cells

Light sensitive cells

Seeing Detail

Before we can practically answer how well the eye can see detail in a scene, let's look at the modern-day digital camera, which is a somewhat simpler device.

Seeing Detail with a Camera

A camera's image sensor typically contains millions of tiny "pixels" or photoreceptors. A high quality camera may have fifteen million receptors, usually quoted as a 15 megapixel camera.

Many other factors can affect image quality, but the total number of pixels defines an upper limit of picture quality attainable.

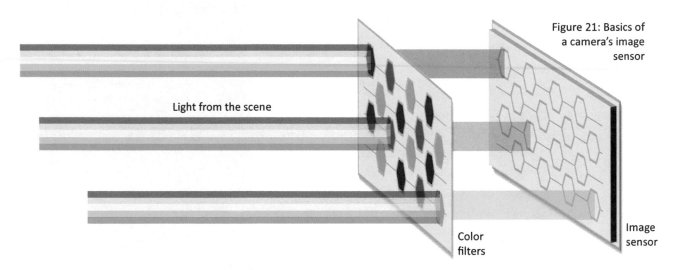

Light from the scene

Color filters

Image sensor

Figure 21: Basics of a camera's image sensor

Typically the camera's photoreceptors all have the same light sensitivity (they are color-blind), but they are overlaid with color filters to make them sensitive to light in the red (R), green (G) or blue (B) parts of the visible spectrum, as shown in the figure.

The camera's photoreceptors are uniformly spaced over the area of the image sensor, so that the resolution, or detail, in the image is the same over the whole picture, wherever we look.

The number of pixels then basically determines how sharp a picture we can obtain.

Returning to our sample scene, let's remind ourselves that the picture produced by our digital camera, with its 6 million color photoreceptors or pixels, is clearly in full detail and full color *uniformly* over the whole picture.

Figure 22: The camera's uniformly detailed, colorful image of a real scene

Seeing Detail with the Eye

Like the camera, the eye broadly analyzes and filters the light into the red, green and blue parts of the visible spectrum.

However, unlike a camera with its color filters, the eye has three separate types of color photoreceptors, sensitive to light in the red (R), green (G) and blue (B) parts of the visible spectrum.

These operate in daylight vision, when light levels are relatively high. The receptors are called cones due to their physical shape.

They total about 6 million, or about the same number as in the digital camera we used to take that detailed picture we showed earlier.

In addition, the eye has some 120 million receptors, called rods due to their physical shape, with only black-and-white sensitivity, used primarily for nighttime vision, when light levels are low.

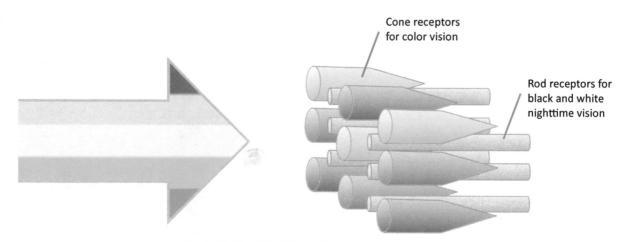

Cone receptors
for color vision

Rod receptors for
black and white
nighttime vision

Figure 23: Simplified illustration
of the eye's image sensor

Since we're dealing with a daylight scene, let's first give our attention to the color photoreceptors.

What we find is that the eye's receptors are not at all evenly spaced over the image sensor's surface - unlike the camera, with its very structured, uniform array of receptors.

In fact, the eye has a huge concentration of color receptors in a very small dish-shaped area, called the fovea, and then the color receptor density rapidly falls off as we move away.

The fovea is located a little way off from the blind spot, and approximately in line with the lens center or visual axis, as can be seen in the illustration.

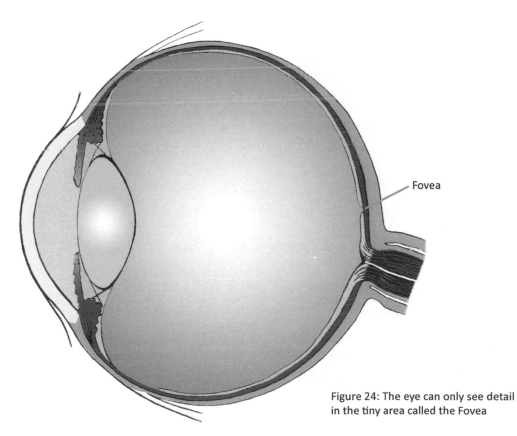

Fovea

Figure 24: The eye can only see detail in the tiny area called the Fovea

So, in reality, at any given moment, we can only see detail over that small area where we have a high density of color receptors - and the rest of the image is a blur.

In addition, to give us the best full-color and detail capability in daylight, Nature has packed the inner part of the fovea, called the foveola, uniquely with our bright daylight color receptors (no rod-based black and white receptors here) .

Over this fovea is a layer call the macula that is yellowish in color.

As an aside, you may have heard of macular degeneration which is a medical condition usually affecting older adults and can lead to severe loss of visual detail precisely because it impacts this "super sensor" area of the eye.

If we hugely magnify this "super sensor" area we can get an idea of the relative size of each part.

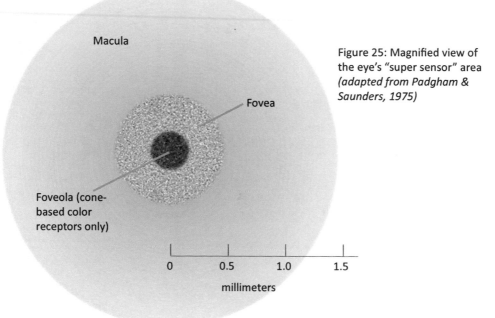

Macula

Fovea

Foveola (cone-based color receptors only)

0   0.5   1.0   1.5

millimeters

Figure 25: Magnified view of the eye's "super sensor" area *(adapted from Padgham & Saunders, 1975)*

To help you practically appreciate just how small an area of the scene we can truly see in detail, the fovea is only about the size of the circle created by touching the tips of your thumb and index finger together, held at arm's length.

Figure 26: We only see detail over this small area

In fact, according to the American official definition, we are all legally blind just 25 degrees from the center of our detail area. At this point our vision has fallen from a normal 6/6 (20/20 in the USA) to 6/60 (20/200) or poorer.

To put this in practical terms, if you're reading this book at a comfortable reading distance of 50 cm (16 inches) then your vision has fallen to being legally blind less than the height of the page away.

Returning to our colorful scene, let's again assume we are looking at the left hand face with our right eye. We only see detail in that left face because that's where we're focusing our attention, and that is where we have our high density of receptors.

Elsewhere, the picture becomes more and more blurred as we move away from the fovea, because there are fewer and fewer color receptors.

Figure 27: Simulation of the detail the right eye sees instantaneously

Don't believe you can only see detail in that small sweet spot?

Try this simple exercise.

Hold the book at arm's length, just like we did with the blind spot exercise. Close your left eye, and then concentrate with your right eye on the center of the star. No cheating by snatching a quick glance.

Now try to identify which letters of the alphabet are in the colored circles. Even the letters in the smaller circles close in are hard to read.

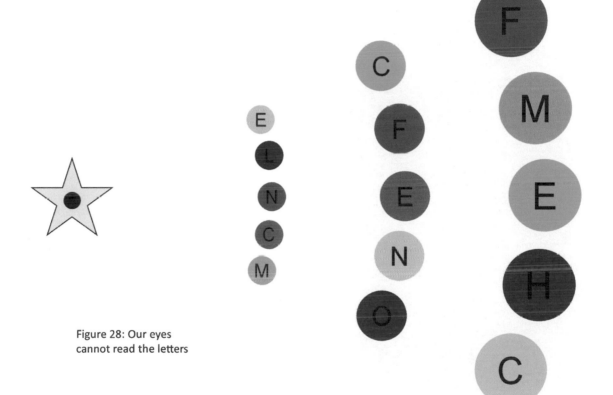

Figure 28: Our eyes cannot read the letters

Move the page nearer to or farther from your eye while concentrating on the center spot in the star. You will be amazed how you still cannot read the letters, even the close-in colored circles.

When attempting to read the letters, you may experience a very strong, automatic urge to move your eye to focus on the letters. This simply illustrates the fact that the eye recognizes its inability to see clearly off-center, and fights

37

to move the eye to wherever it is you really wish to see in detail.

As an aside, try again concentrating on the star and then identifying the various colors used. Even this is not easy. Although you do get an *impression* of colors, it is very hard to say what colors, and where they are located.

Here's another simple exercise to try with a friend.

Get a pack of playing cards. Ask your friend to look intently at a spot on a wall and not to move her gaze. Meanwhile, stand by the wall, a few feet to one side of where she is looking. Hold up any one card at a time and ask her to guess the suit and number of the card.

She will be amazed that she can't do it, even though the card is so close.

Figure 29: Playing card demonstration

While still holding the card up, slowly move along the wall towards the spot where she is looking and notice just how close you have to be to that spot before she can truly identify the card color and number.

* * *

From this simple test, it is clearly biologically impossible for anyone to "photographically" capture a whole scene or a whole page of text instantaneously - despite many claims to the contrary.

So-called speed reading is achieved by training the eye to better move rapidly from a block of a few words, to the next block, and so on. We will cover this further when we explain eye movements.

The Eye's Megapixel Rating

People are so used to rating a digital camera by the number of megapixels in the image sensor that the temptation is to ask what the eye's megapixel rating is.

I hope you can now see that it is not easy to compare the megapixel rating of a camera, with its nicely uniform image sensor, to that of the eye with its very small "super sensor" area and two sets of quite different sensors - daytime color and nighttime black-and-white.

But we explained earlier that the eye has around 6 million color receptors, so isn't the rating just 6 megapixels?

No, because, as we have now seen, the vast majority of color receptors are concentrated in a tiny area, giving us much sharper vision there, and much poorer vision in the periphery.

So it is not very meaningful to put a megapixel rating to the eye.

However, we can put a number to the eye, if we assume a camera must be able to take a picture that provides similar quality, *over the whole picture*, to the eye's best spot capability, on the basis that the eye could look anywhere over the camera's picture.

We would then need a camera with an image sensor of over 500[1] megapixels.

Why a Camera has a Shutter but not the Eye

Basically a camera has a shutter to block the light entering, until we are ready to take a photo. Then it opens temporarily to allow just the right amount of light onto the image sensor, and provide a picture that is neither too dark (under-exposed) nor too light (over-exposed).

A second function of the shutter - less well understood perhaps - is to avoid, or at least minimize, blur in the image.

I say minimize because we are all familiar with using a camera and the disappointment of pictures that are blurred, or out of focus.

So what causes blur in a camera, even though it has a shutter, and how does the eye typically *not* perceive blur, even though it doesn't have a shutter?

Let's look first at the camera and the role of the shutter.

The Camera Shutter

You may recall from an earlier chapter that the camera's image sensor is normally covered up from exposure to light by the presence of the shutter, which is a mechanical or electronic "blind".

When you take a photo, the shutter opens briefly and allows light from the scene to fall on the image sensor. Normally, the shutter is only open for a very small fraction of a second - unless you're taking night shots.

41

However, as we have all experienced, even an automatic camera with image stabilizers can produce disappointingly blurred photos. Here are four key categorizations showing why camera blur still occurs, and their causes.

1. The whole picture is blurred

We didn't hold the camera steady enough while taking the shot - perhaps because the light level was low and the exposure time long, so our natural hand movement caused the image to wobble on the image sensor.

Figure 30: Overall blur due to low light level and camera shake

Many cameras today have built-in image stabilization which attempts to counteract hand shake and so at least reduce blur due to longer exposures.

2. Only some items are blurred

One or more items in the field of view moved, or was moving, while the shutter was open.

In the example below, the car is moving, and so blurred, while the storefront remains stable and sharp.

Figure 31: Partial blur due to some items moving during exposure

Holding the camera as still as possible, or using image stabilization, *cannot* counteract this type of partial image blur.

As a photographer or movie-maker, in this situation, you must either keep the camera still and let moving objects blur, or conversely, make the camera track the moving object and then have the background blur. This latter technique is referred to as panning and is often used to illustrate movement - especially in photographing sports. Here is an example:

Figure 32: Blur of surround due to following a moving item
(© *Matthias Trischler, 2009*)

3. Near items are blurred

This is similar to the previous category, since we see partial blur in the image. However, in this case, blur is greatest for near items while distant ones are in sharp focus.

It arises when trying to take a picture of a scene while you're holding the camera perfectly still, but the camera (and you!) are moving relative to the scene - for example, taking a picture from a car or train.

Notice in the example below that the nearby tree and bushes are severely motion blurred, while the somewhat more distant building is reasonably well defined.

Figure 33: Near blur from a traveling camera

Again, holding the camera as still as possible, or using image stabilization cannot counteract this type of partial image blur.

* * *

Normally, a camera tries to take a picture as quickly as possible, with the shutter opening and closing in a fraction of a second.

The camera attempts to "freeze" the scene and so avoid any significant movement that would show up as blur.

An important point here is that, in all of the above examples, blur could have been minimized if we could have used a faster exposure.

Taking a picture as quickly as possible means that the camera's aperture, or lens opening, must be large - to capture the necessary amount of light during the short time the shutter is open.

Unfortunately, having a very large aperture exacerbates another category of blur, described next.

4. Items nearer or further away are blurred

Blur arises here because, in reality, there is only one distance that is precisely in focus. All others are nearer or farther away from the perfect focal setting, and so are more or less out of focus.

The distance over which the change of focus is *not* noticeable is referred to as depth of field, and high quality camera lenses will often mark the estimated depth of field at each aperture setting.

Again, holding the camera as still as possible, or using image stabilization *cannot* counteract this type of partial image blur.

Of course, like panning your subject, you might use this technique deliberately to achieve the very nice effect of an item standing out in sharp focus while its surround is blurred. Here is a good example of this type of blur.

Figure 34: Blur from limited depth of field

* * *

In summary, we have identified four categories of blur arising from a camera, and only the first type can be reduced by even the most sophisticated image stabilizers.

This helps to explain why you can still get blurred pictures even with a very expensive, and technically advanced camera.

47

The Shutterless Eye

We have seen that the camera requires a shutter to properly expose the scene onto its image sensor, and at the same time minimize the different types of blur that can occur in a photograph.

On the other hand, we have stated that the eye has no shutter (remember the eyelids are there for lubrication, not to act as a shutter). And yet the eye doesn't normally perceive blur. How can this be?

Since the eye doesn't have a physical shutter, its image sensor is continually bathed in light from the scene. As we move around, or simply move our eyes, the image on the back of the eye is continually dancing about.

At first sight (pun intended!) Nature's approach would appear to be a recipe for disastrously blurred images.

Imagine what it would be like if you set your modern-day advanced camera to a long exposure, even for say 10 seconds, and then just gaily moved the camera around for the duration of the exposure. What you would have as a still photo would be completely blurred and unintelligible.

Similarly, if you casually moved around with a movie camera, in the same way our eyes move around, you would get all sorts of blur - along with a strong sense of motion sickness.

So we understand that, except for special effects, it is very important to keep a camera still, relative to the item of interest in the scene, while taking a picture, to the extent of using a tripod when necessary.

Strange as it may seem, the eye follows a quite contrarian approach. It is continually and involuntarily vibrating, even when you attempt to view a scene steadily.

These *tremors* avoid individual receptors in the light sensitive area from becoming saturated or fatigued with continuous exposure to a steady level and pattern of light from the scene being viewed.

In fact, if you use a special apparatus to make the scene appear stationary on the eye, after just a few seconds the image will simply fade to gray.

Here is an easy exercise that demonstrates image fading.

Figure 35: Image fading - the Troxler effect

Hold the book at normal reading distance and concentrate with both eyes on the center spot. After a few seconds the pink ring will start to fade, and eventually disappear into the pale blue background.

As soon as you move your eyes, you will see the ring immediately reappear.

So, contrary to the use of image stabilizers that are now a key feature on better cameras to minimize the blurring effects of camera shake, you can see that their use on the eye would be completely counter-productive, and simply result in a gentle fade to gray.

Essentially what is happening is that the receptors become chemically tired and no longer respond to light from the scene.

Similar to image fading, it is easy to demonstrate saturation of the receptors.

Hold the book at normal reading distance and concentrate with both eyes on the gray dot in the center of the rather oddly colored yellow and cyan flag. Concentrate on the dot, while counting slowly to 15. Try not to blink.

Now move your eyes to focus on the black dot below the flag. You will immediately see a properly-colored, red-white-and-blue Union Jack flag for Great Britain.

As the individual receptors tire, they no longer respond well to their color. That area of the eye then becomes less sensitive to that color, and so we are more aware of the opposing, or complementary color.

The effect will not last more than a few seconds, as the individual receptors start to recover. The brighter the light you try this under, the better.

Figure 36: Afterimage demonstration of receptor saturation

When we look at the oddly colored flag, we reduce our sensitivity to yellow, leaving us more sensitive to blue, so those yellow parts now look blue on a white sheet of paper.

Similarly, the cyan color reduces our sensitivity to that color, and reinforces our sensitivity then to its opposing color, red. The black areas don't saturate our vision, so in those areas we see white.

When we say opposing color, we simply mean that if we reduce that original color in the visible spectrum of colors, then what remains takes on the opposite tinge. We will explore this further when we cover how we see color.

A modern camera avoids this saturation by assessing the amount of light in the scene, and then opening the shutter and leaving it open only for the amount of time necessary for the image sensor to capture enough light to obtain a good picture (not too light and not too dark), and of course as fast as possible to avoid blur.

The shutter is then closed and no more light allowed into the camera. The image meanwhile is swept off the light sensitive area into the camera's memory, ready for the camera to take another picture.

In summary, unlike a camera which has a shutter, the eye must make continual, small movements simply to avoid its receptors saturating or fatiguing.

A huge advantage of NOT having a shutter is that the eye is continuously open and ready to detect danger in the real world around us. Of course, we still haven't explained why the eye doesn't normally see blur - especially now we know the eye moves around continuously.

We will explore this apparent incongruity in the next section.

Eye Movement

We learnt earlier that, at any one moment, the eye only sees detail over a very small area, although we perceive life around us in apparent full, rich color and detail.

So, in addition to the fast, small tremors and slow drifts, the eye must also be jumping from one area in the scene to the next, and the brain must be piecing together all those highly detailed snapshots of the scene to build what we perceive as an apparently complete image in full, rich detail.

Figure 37: The eye absorbs detail in a small area
then jumps to the next area and so on

This is exactly what happens in fact. The eye is continually moving around, taking in the scene through "bite-sized chunks".

It does this not smoothly, but in jumps, called saccades. At the end of each jump, the eye fixates on a spot, absorbs the detail in that area, then jumps again and so on. This is illustrated in the previous figure where we have notionally divided the scene into jigsaw puzzle pieces.

The eye and brain build up a perceptual whole from all these "jigsaw" puzzle pieces.

That is not to say that we create a super high definition picture in our heads. Instead the eye and brain analyze the contents of the picture and identify its key attributes.

We don't see blur because we are quite blind each and every time the eye jumps to a new part of the scene. The eye and brain work together to simply blank out our vision during the jump - we are temporarily blind every time our eyes move.

Normally we have no perception of this temporal blindness, but it is very easy to demonstrate.

Try this for yourself: look closely at your eyes in a mirror. Now, with both eyes open, glance first at your left eye then your right eye. Repeat this, looking quickly from one eye to the next.

You will not observe any movement of your eyes, nor any movement blur.

The eye and brain have cleverly contrived to suppress the movement, and hide the fact that there even was movement. This temporal blindness is one of many distraction tricks used by magicians.

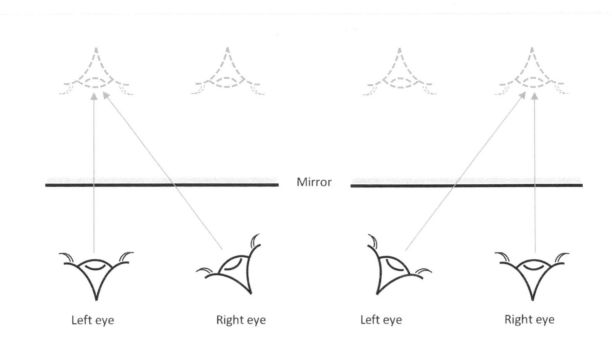

| Left eye | Right eye | | Left eye | Right eye |

Figure 38: Can you see your own eyes move?

Now get a friend to watch you - they will of course see your eyes swinging quickly back and forth.

Finally, swap roles and see in your friend's eyes what you couldn't see for yourself.

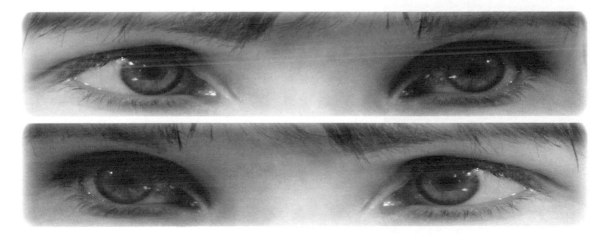

Figure 39: But can you see their eyes move?

Reading and the Eye

So if we can really only see detail over such a small area, and need to make multiple jumps to absorb one page, how do we read a whole page of text?

Reading is a special case of these saccades or jumps, where we focus for a short time on a few words of a sentence (fixation), then jump to the next few words, and so on. The fixation time is typically about a quarter of a second.

In the West, we're taught from an early age to read left to right, one word at a time, and then to go back to the left and start afresh with the next line.

Behind the scenes, just like in the previous illustration of how we take in a pictorial scene, we are actually jumping, fixating, then absorbing just a few words at a time.

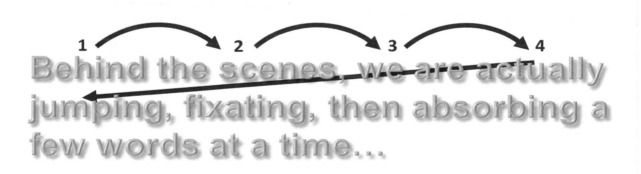

Figure 40: How we read

We have to do this because we only properly see about one small word at time in full detail.

At first sight, this seems almost unbelievable, but the reality of what we can read is illustrated below:

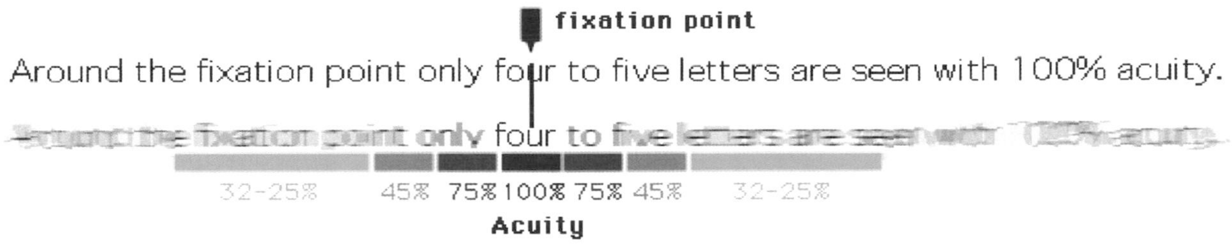

fixation point

Around the fixation point only four to five letters are seen with 100% acuity.

32-25%   45% 75% 100% 75% 45%   32-25%

Acuity

Figure 41: Our instantaneous reading ability
(Hans-Werner, 2008)

So is speed reading a myth?

No, because in practice, when we jump and fixate along a line of text, we often go back and re-read a block, or not move cleanly from one jump to the next.

Speed reading techniques increase reading speed and efficiency by helping to train the eye to absorb the content better the first time, and by training the person to make jumps more consistently in one direction without going back so often.

You can see how different languages, requiring up-down or right-left reading, are simply acquired skills - although of course it's much harder to learn once we are adult.

* * *

Given the fact that at any one time we only see detail over a very small area, and that we are blind during the jump to the next area, so, you may ask, why don't scientists employ some of these techniques for television and movies today?

57

In fact, several attempts have been made to mimic the eye's "super sensor" approach and present detail only in the area where the person is looking, while blurring the rest of the image.

This has met with limited success - first because it is only useful for one person viewing at a time, and second, because it is extremely complex to identify where the eye is looking and instantly provide detail only in that area.

Today, we are able to store and transmit images at low cost and high speed, so there is much less demand for minimizing the amount of information in pictures.

Summary Comparison

While initially appearing quite similar, we can already see fundamental differences between the eye and the modern-day digital camera.

Far from being a near-perfect instrument, we have illustrated the following key limitations of the eye:

- Each eye has a large blind spot where the optic nerve carries signals to the brain

- Each eye has only a small area over which we can see detail *(super sensor)*

- The eye has two different sets of receptors - a color set for daytime, and a black-and-white set for nighttime or dim-light vision

- The eye has to look at the world through the "spider's web" of all the optic nerve fibers, blood vessels and pre-processing neurons

- The eye doesn't have a physical shutter like a camera, other than occasional eyelid blinking which only serves to lubricate the outside of the eye, not to cut off light like a shutter does for a camera

- We are continually, temporarily blind as our eyes jump and then fixate on items of interest in the scene - even though the eye is fully open.

Perhaps our ability to see detail in one small area at a time should not be considered a limitation at all, but a very clever biological approach to provide the finest vision possible, without burdening us with an unmanageably huge megapixel video camera in each eye.

Given this constraint, then, the jump-and-fixate approach to capture all the detail in a scene appears to be a natural consequence.

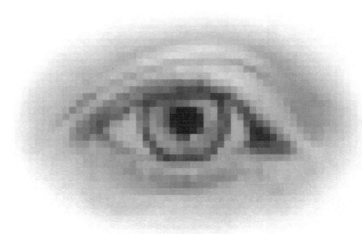

Focus in the Camera and the Eye

This chapter looks at how the camera and eye employ very different techniques to focus on a scene and so obtain the best possible sharpness.

We also cover the eye's limitations and defects that lead to many of us having to wear glasses or contact lenses - if not now, then later as we age.

How the Camera Focuses

We mentioned earlier that the *camera obscura* traditionally just has a pinhole opening, and no lens. So how does it optimally focus?

In fact, by having just a pinhole and not a wide aperture, light from near and far items in the scene are usually in reasonable focus. This, however, is at the expense of image brightness, and images from a *camera obscura* are often disappointingly dull except on bright, sunny days.

Low-cost and disposable cameras typically rely on a fixed wide-angle lens with a relatively small aperture to ensure that all items from about 3m (10 ft) to infinity are in reasonable focus.

Most cameras, though, have adjustable focus - the basic action of which is shown in the next sequence of figures.

A camera lens has a fixed, light-bending power so adjustable focus cameras rely on moving the lens toward and away from the image sensor to achieve optimum focus on the primary item of interest in the scene.

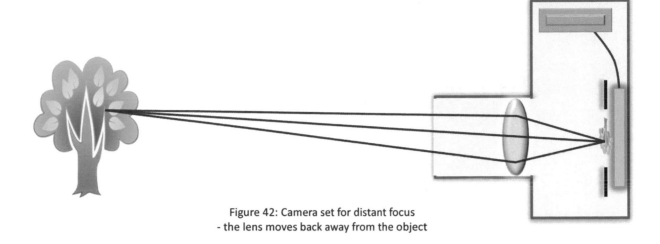

Figure 42: Camera set for distant focus
- the lens moves back away from the object

This can be manual, or more typically today, fully automatic as you press the shutter, with various auto-focus
modes and speeds, depending on the scene type and
content.

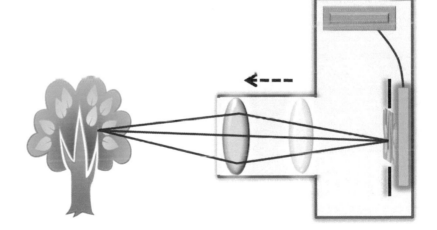

Figure 43: Camera set for
near focus - the lens moves
forward towards the object

A classic auto-focus challenge is that of taking a picture
of two people side by side. You don't want the camera
to focus on the area in the center which may be very far
away, and leave the people blurred.

61

As you press the shutter, a modern camera will typically show you rectangles in the viewfinder where it is attempting to focus - making a best estimate at what you are intending to capture.

How the Eye Focuses

In the eye, light is first bent at the cornea - the interface between the air and the fluid inside the eye. The lens inside the human eye completes the focusing.

As we saw from the eye cross-section, this lens is fixed in position within the eyeball and so cannot move back and forward, like a camera lens, to focus.

Instead, Nature has endowed us with a lens that can change shape and so change its light-bending power.

Ring-shaped muscles around the lens compress or relax to achieve this change.

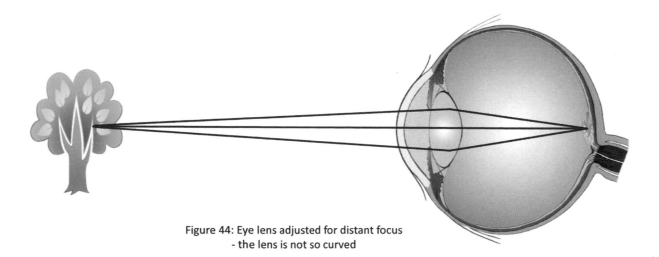

Figure 44: Eye lens adjusted for distant focus
- the lens is not so curved

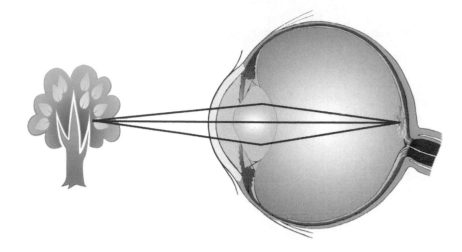

Figure 45: Eye lens adjusted for near focus - the lens is much more curved

As an aside, while humans and primates generally focus by changing the shape of the lens, cephalopods (octopus, squid etc) change their focus by moving[2] their lens back and forward - much like a man-made camera.

Why We Need Glasses

In general, the eye's focusing can be incorrect for two reasons - either the curvature of the cornea or the length (depth) of the eye is inappropriate to focus the image correctly on the back of the eye.

If we're near-sighted, the cornea is curved too much, or the eyeball is too long, and the image is focused in front of the inside surface of the eye. This is called myopia and can be corrected by an external concave lens (either as glasses or a contact lens) as shown in the next figure.

Conversely, if we're far-sighted, the cornea is curved too little, or the eyeball is too short (not deep enough), and the image is focused beyond the inside surface of the eye. This is called hyperopia and can be corrected by an

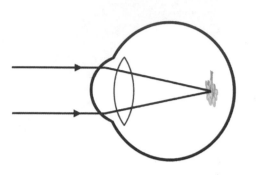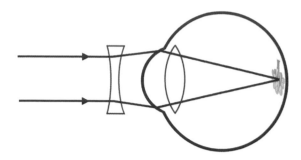

Figure 46: Near sight (Myopia) and its correction

external convex lens (either as glasses or a contact lens), as shown in the next figure.

Instead of having glasses or contact lenses, and since we cannot easily change the length of the eyeball, we can correct for the above errors by artificially changing the shape and so the bending power of the cornea.

Corneal procedures include LASEK and PRK, which are somewhat similar, and LASIK which is a little different.

As we get older, the lens ages and becomes less flexible - less able to make itself into a high power lens for close viewing. This typically arises from about age 40, and all normally corrected eyes will need reading glasses as we age.

The term for this is "old sight" or presbyopia. If we haven't yet experienced it ourselves, we have seen people not able to read close up, and resorting to holding their cellphones or newspaper farther and farther away.

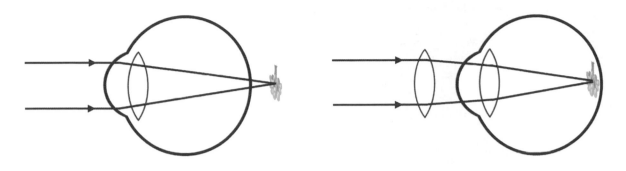

Figure 47: Far sight (Hyperopia) and its correction

The lower the light level, the more the eye's pupil has to open to allow more light in, and the worse the effect of "old sight". That is why it is even harder to read a menu in a dimly lit restaurant.

Clearly, remodeling the cornea, as in LASIK or PRK surgery, cannot directly help this "old-sight" problem. Unfortunately, to avoid reading glasses some surgeons recommend remodeling each eye differently - so-called mono-vision - with one eye corrected for near and one for far distances.

With only one eye now able to see well in the distance, it is questionable whether this is an appropriate treatment since overall distance vision is somewhat impacted. There is also some small reduction in three-dimensional depth perception. This can slightly impact handling of nearby objects, especially as one ages and cannot focus close-up as well as previously.

As an aside, a cataract is simply a clouding of the lens inside the eye that can also occur with aging. Here, modern surgery can readily remove and replace the cloudy lens with a suitable synthetic one.

Exposure in the Camera and the Eye

Photographers and film makers adjust the lens opening, termed the aperture, and/or the shutter opening time to obtain a correctly exposed picture - not too dark, and not too light.

(a) Wide or large aperture   (b) Small aperture

Figure 48: A camera's aperture set open (a) wide and (b) small
(adapted from Mohylek, 2006)

With early cameras it was just a guessing game as to how long to expose the film to get a good picture.

Later, separate light meters became available. These were used prior to taking the picture, with the photographer having to walk over to the scene of interest and take a reading. The reading was then translated into an aperture and shutter speed setting - obviously a challenge if light in the scene changed in the meantime.

Today of course the light meter is integrated into the camera body and in "point-and-shoots" may be fully automatic, controlling aperture and shutter speed together.

The camera still guesses at what the photographer is wanting to take - from a wide aperture, with fast shutter speed for moving objects, to a small aperture, with slow shutter speed to improve depth of field.

We know the eye has no shutter, but it does have a pupil which is the opening created by the iris, like the aperture adjustment on the camera. Is this all the eye has then to adjust exposure?

(a) Large pupil   (b) Small pupil

Figure 49: The eye's aperture or iris open (a) wide and (b) small
(adapted from Nicko Va, 2009)

The answer is no. In addition to the iris, the eye has a very smart system to be able to see over light levels that range from extremely bright sunlight, to almost total darkness. In fact, the iris is an imperfect aperture control since it may be affected by factors other than strict light levels - stress and drugs being examples.

While the camera's receptors rely on physics and electronics to provide photosensitivity, the eye's receptors, on the other hand, rely on chemical dyes that "bleach" according to the amount of light received.

For each receptor, molecules that bleach will, over time, regenerate, dependent on the amount of light received in subsequent scenes. So the eye's image sensor has the ability to continually "adapt" to the light levels in the scene.

In addition, to cope with the vast range of light levels encountered in the natural world, Nature has endowed our eyes with two intertwined image sensors - one comprising cones for color vision in daylight and the other comprising rods for black-and-white vision in dim light.

In daylight, the rod receptors are in a state of continual saturation or bleaching. As you enter a cinema, they slowly regenerate in the dim light and give you low light vision - albeit in black and white. Meanwhile the screen brightness is sufficiently high to continue to stimulate the cone receptors so you see the movie in glorious color (and often 3D today).

In practice, this auto-adaptation takes time - as you will have experienced in going into a cinema from broad daylight. Initially you cannot see inside the theater at all, and then slowly, over several minutes, you start to make out enough of an outline to be able to move around safely, without continually colliding with people and chairs.

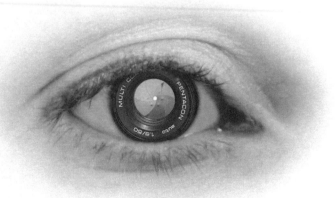

Powering the Camera and Eye

The camera, of course, is powered by a battery, enclosed within the camera body. This battery powers the electronics (the image sensor, the processor and memory) as well as the electro-mechanical items that operate the aperture and focus.

The eye is essentially powered through the arteries and veins.

If we look through the pupil into the eye, what we primarily see are the retinal blood vessels, and not the image sensor with its millions of photoreceptors.

This is illustrated below, where we also show the location of the blind spot and the "super sensor" area called the fovea.

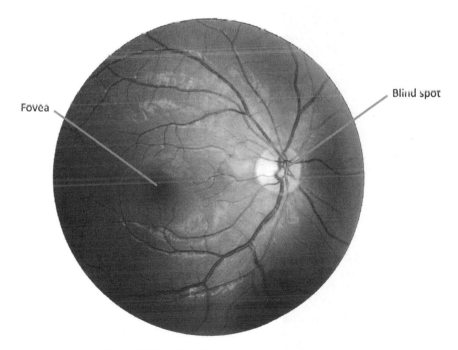

Figure 50: View into the eye showing the back inside surface and the retinal blood vessels
(adapted from Ignis, 2007)

We see the retinal blood vessels when we look into the eye because, as noted earlier, they are actually in front of the eye's image sensor, not behind or between.

You can observe these blood vessels by closing your eyes and shining a modern small LED flashlight close to the lower right corner of your right eye (or reverse this if you prefer your left eye). Now shake the flashlight quickly from side to side. You will see the blood vessels as brighter lines emanating from the blind spot.

As an aside, we have all experienced the dreaded "red-eye" when we take photos of people (or animals!) with a flash, as shown here with a lovely pair of rabbits.

Figure 51: Example of the red eye effect from a camera flash
(PlePle2000, 2006)

Light from the flash enters the eye through the pupil and some reflects back towards the camera. The eye's image sensor, along with the nerves in front of the sensor, are relatively transparent so the light penetrates to the choroid whose dark-colored pigment absorbs mostly green and blue light, so reflecting red light back to the camera.

The choroid sits behind the eye's image sensor and provides oxygen and nourishment to the outer layers of the retina. So the reflected light is red not from the retinal blood vessels themselves (as is often stated) but from the choroid. The choroid provides the overall red-orange background in the previous figure showing the inside view of the eye.

To avoid or at least minimize red eye, get the subject to look away from and not directly at the camera, use a separate flash that is well away from the camera so the reflected light doesn't come back towards the camera, or best of all use a bounce flash with reflecting silver umbrella to disperse the flash light source.

Some cameras have a red eye reduction setting which produces a pre-flash to temporarily force the pupil smaller and so reflect less red light back out of the eye.

The eye is the only part of the human body where we can non-invasively see blood vessels in actual operation. This provides a tremendously powerful tool for early diagnosis of some key illnesses.

As an aside, you may have noticed cats' eyes reflect quite differently under flash - usually green, not red like humans. This is due to cats having a special layer behind their image sensor that reflects light broadly in the green part of the spectrum. This helps cats see better in low light - but at the expense of slightly poorer sharpness and daylight vision.

Seeing Brightness

It might be thought that seeing brightness is the simplest and easiest part of vision. After all, light is light and we cannot incorrectly measure the strength of each ray entering our eyes.

In reality, while our eyes do receive the raw intensities of the scene, what we perceive is very dependent on the context of the scene - what currently surrounds the area of interest - locally and overall.

In fact, just knowing the brightness or color of an item is not very useful in the real world with its huge variations in light and color.

To cope with these huge variations, the eye must adjust to global differences, and then seek local differences in brightness and/or color. Such differences are called brightness and color contrast.

Ultimately, it is the contrast between items in a scene that gives us a sense of each item's shape and texture, and enables us to label them appropriately.

A sheet of white paper appears white both in bright sunlight, and when we see it under dim light. Similarly a black item in the same scene continues to appear black, even though under bright sunlight it may in fact be reflecting much more light than the sheet of white paper reflected under dim light. It is the *relative* brightness of items in a scene that matters - so-called brightness constancy.

Because the eye is now making judgments based on the scene content rather than absolute intensities, the eye can be fooled by specially designed patterns.

Here we have a flock of gray birds flying off into the darkness. Every gray bird is identical in size and color, but because the background gets darker as we move eastward they appear to become lighter and lighter as they approach the night.

Figure 52: Flying birds are all the same gray

This type of illusion has been used over the centuries to great effect in Japanese art - especially in paintings and porcelain.

A dramatic example of how local brightness contrast can fool the eye can be seen in this next illusion.

Figure 53: Squares A and B are the same
(Adelson, 1995)

The squares labeled A and B are of the same intensity. If we place a light meter over each square we get the same reading.

Yet the eye, even though it now knows them to be the same, cannot override the impact of local contrast, and continues to interpret the two squares differently.

Let's crop the scene down to include just the two squares A and B.

Figure 54: Do the squares A and B still appear different?

Still the illusion is very strong. We can crop huge areas of the scene and still not be able to comprehend that A and B are the same brightness - even though we know they are.

To illustrate they are in fact the same, in the next figure I have simply bridged the two squares with a block of the same gray.

Figure 55: Squares A and B really are the same

Now the eye sees the two squares as the same or very nearly the same. If we removed all the other local contrast clues they would appear identical.

The illusion is particularly strong because we assume, reasonably, that the pattern is uniformly comprised of light and dark squares, even though the actual light intensities are quite different where the shadow applies.

Then we subjectively correct for the impact of the soft-edge, life-like shadow created by the column. This shadow quite deliberately only completely shades a few light squares, with adjacent dark squares being in transition from shadow to light.

In practice, if the eye (and brain) didn't subjectively correct for all the shading and texture gradients that occur in the real world, we would have a very hard time identifying objects, and moving around, without colliding with everything in our path.

If we offer up a picture of nicely even steps of gray, we will perceive each step as being slightly shaded. Each step becomes slightly darker as it approaches the adjacent lighter gray step, and vice versa. This phenomenon, known as Mach bands, is illustrated below.

Figure 56: Uniform gray bars seem shaded (Mach bands)

We will explore this effect in more detail later when we discuss how the eye sees patterns and shape.

Seeing Color

The Nature of Light

As is well known, white light comprises all the colors of the rainbow. But what is light exactly?

Light is electromagnetic radiation - or simply another part of the radio spectrum, like the FM radio band.

For humans, we can see over a frequency range of about 400 to 790 terahertz or, in wavelength terms, 750 to 380 nanometers. Frequency is simply the reciprocal of wavelength, scaled by the speed of light in a vacuum.

If we stretch the human visible spectrum out onto a chart, we can see the wavelengths and frequencies that apply to each *perceived* range of color:

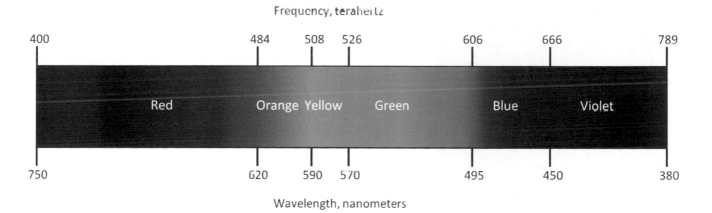

Figure 57: The range of visible light

Comparing this range with the broadcast FM band, light's wavelengths are around five million times smaller.

It is important to recognize that light itself is *not* colored - it is simply our *perception* of the different wavelengths in a scene that gives rise to the sensation of color in the brain.

Another way of thinking of color is that the brain adds labels, which we call colors, to the different wavelengths - or combinations of wavelengths - received.

Color Vision

As we briefly described earlier, both the camera and the eye use an image sensor that is sensitive to light in the red, green and blue parts of the visible spectrum. This is all we have to represent the entire color gamut in a scene.

Each receptor is broadly "tuned" to a different wavelength, or frequency, much like you tune into a radio station on the FM band. Since light itself has no color, in reality it is better to describe the three receptors as being tuned to long (L or red), medium (M or green), or short (S or blue) wavelengths.

The relative sensitivity of each of the eye's three receptors is shown in the next figure.

Again, note that light is not colored, and that the "color" receptors are not colored. In fact the photo-sensitive chemicals in the color receptors are transparent. Their chemical structure simply enables them to absorb light preferentially over the visible spectrum and produce an electronic signal based on the energy absorbed.

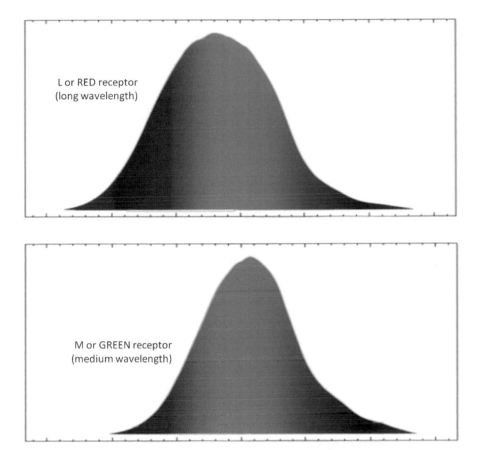

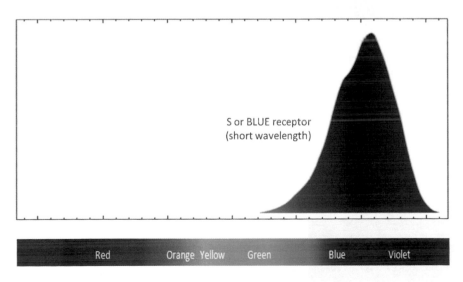

Figure 58: How our three color receptors respond to light
(adapted from Stockman and Sharpe, 2000)

You'll notice that the "red" or "L" receptor peaks in the orange-yellow (rather than red) and is very broadly tuned - seeing light in the red, green and blue parts of the visible spectrum.

The "green" or "M" receptor peaks in the green, but also sees light in the red and blue parts of the visible spectrum, although not as broadly as the "red" receptor.

Finally, the "blue" or "S" receptor mostly responds to light in the blue region, together with a little green.

These broad-based sensitivities help to minimize color illusions, or errors, when identifying the vast array of possible colors.

So light from each part of the scene is received by the three receptor types, and the signal from each receptor simply indicates how much energy in the frequency range is received for that receptor, in that part of the scene.

Taking as example our usual detailed, colorful scene, we can zoom in to two differently colored areas and illustrate typical outputs from the three color receptors.

This first analysis by the eye is illustrated in the figure opposite.

By analyzing the light coming from the scene, using our three color receptors, we can adequately represent most colors in the visible spectrum.

It has been estimated that the eye can see approximately ten million[3] different color shades.

Figure 59: How our three color receptors respond to colors in a scene

The three so-called primary colors of red, green and blue are termed *additive* - more red light from the scene adds to the existing light and we then perceive the light with a red tinge.

If all three colors, red, green and blue, are present then we achieve white light. Televisions and computer screens work through additive color mixing. See the top illustration opposite.

This is in contrast to how colors are used in painting or printing. Here, in the presence of white illumination, the *absence* of any paint or ink on the canvas reflects white light.

A cyan paint absorbs red, but reflects blue and green light.

A yellow paint absorbs blue, but reflects red and green (which as we show later, is seen by the eye as yellow).

If we overlay these two paints, then both red and blue are absorbed, leaving only green to be reflected.

Each paint absorbs or subtracts from the white illumination, and so mixing paints or inks is termed subtractive color.

The typical primaries used in painting and printing then are cyan, magenta, and yellow. See the illustration opposite.

With just three color receptor types, there are limitations in the range of colors that can be perceived, along with some color illusions that arise. These color illusions can be thought of as simply errors or confusions arising from having only three color receptor types.

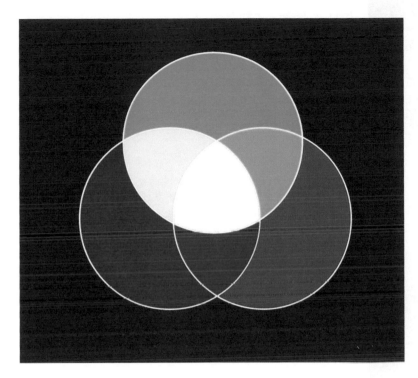

Additive Color Mixing

- Start with a blank screen that in the absence of light sources produces black

- Red, Green and Blue light sources

- When all three colored sources are active together they combine to produce white light

- Television and computer screens

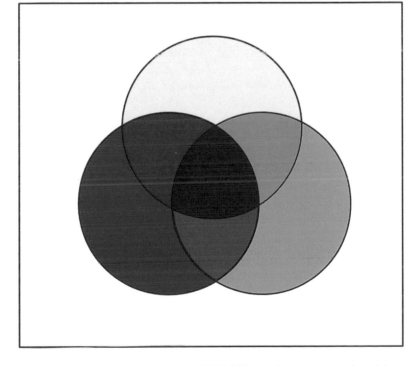

Subtractive Color Mixing

- Start with a blank canvas or sheet of paper illuminated by white light and so the absence of inks produces white light

- Cyan, Yellow and Magenta inks or paints

- When all three colored inks are present together they combine to absorb the white light and so produce black

- Painting and printing

Figure 60: Additive and subtractive color mixing

The beauty of the eye is that it is well adapted to sufficiently meet the needs of surviving and living safely in the real world, with its natural illumination.

Real world errors are uncommon, but special colored patterns that rarely or never occur in nature can be shown to "trick" the eye, and so create an illusion.

A simple example of an illusion relates to the color yellow. Let's say we have a "true" yellow whose wavelength is around 580 nm. Both the red and the green receptors will respond, giving us the sensation of yellow.

Now let's take two different spectral colors, one in the green part and one in the red part of the spectrum, but no yellow at all. Again, both the red and the green receptors will respond - and give us the same, but now illusory, sensation of yellow.

So the human eye - and the camera - cannot distinguish between a mixture of red and green, and a "true" yellow.

The illusion is a useful one for eye doctors who use an instrument called an anomaloscope that presents "pure" monochromatic yellow in one half of a display, and a red-green mixture in the other half, to assess abnormalities in a person's red and green color vision.

Unfortunately, we cannot truly demonstrate this on screen or in print because these man-made systems suffer from similar limitations. However, here is a simple simulation of the anomaloscope.

Prop the book up so you can readily see the picture. Now move away while looking at the picture, until the right hand set of small red and green squares blurs together. You will now see a yellow that approximates the yellow on the left hand side.

Figure 61: True yellow and a red-green mix simulation

Color Blindness

In the United States, about 1 in 13, or about 8%, of the male[4] population either cannot distinguish red from green, or see red and green differently.

Clearly with this level of problem it is important to design color coded items to be distinguishable in some additional manner, and not just use color.

For example, traffic lights use red and green to indicate stop and go - and obviously it is very important to make sure that the 8% of the male population doesn't get it the wrong way around.

Conventionally, the lights are arranged in a vertical pattern with red at the top so the position of the light, not just its color, can be used for identification.

Unfortunately, we occasionally see traffic lights placed horizontally. Should the red light be on the right or the left?

Figure 62: Horizontal traffic lights

The color defect situation is much less dramatic for the female[4] population where only about 1 in 250 or 0.4%, is affected.

This stark difference arises because color defects are typically inherited from mutations of the X chromosome.

The male population inherits only one X chromosome, while the female population inherits two. If a female has one good X chromosome this will override the mutated chromosome and she will not be color defective.

* * *

Initial testing for color vision defects uses the well known Ishihara charts. Below is an example.

The number "74" should be clearly visible to persons with normal color vision. Persons who are dichromat (only 2 receptors functioning) or anomalous trichromat (a catchall term for having all 3 receptors, but with some abnormal sensitivities) may read it as "21", and viewers with achromat (no color vision) may see nothing.

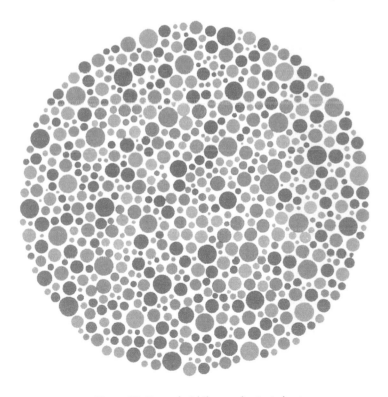

Figure 63: Example Ishihara color test chart

Lighting and Color

Another simple example of how the eye can be tricked because it uses only three receptors to see all colors is that white light can be perceived using just two "true" (monochromatic or single-wavelength) colors - a blue, together with a yellow (which, as we have seen, stimulates both the red and green receptors).

Neither of the example illusions is particularly a problem in real life. However here is an example of a real world problem even for those with normal vision.

You have probably encountered the disappointment of buying apparently matching clothes in a store, only to find on getting home they look a very different color in daylight.

This is a well known phenomenon called Metamerism. In this case, the color you perceive is affected by the very different nature of "white' light used in stores (often fluorescent) compared with that of natural daylight.

Fluorescent lights, and the newer CFLs now being used in homes, all suffer to a larger or smaller extent from the problem of emitting light that is "spiky" in nature compared with the continuous spectrum of daylight, or light from a regular incandescent bulb. This is why there is more chance of a metameric illusion arising with fluorescents.

Cameras try to correct for the different types of illumination but can only do so to a certain extent. That is why, for example, under incandescent light, your photo or video may have a yellow-orange tinge or color cast. The eye rarely suffers from this problem, being able to well adapt. We explore this topic further when we discuss color constancy.

The spiky nature of fluorescent light exacerbates the problem - causing some items to apparently shift color compared with what the eye sees under normal daylight.

The chart shows a sample fluorescent tube light output (blue curve) compared with the standardized output of a regular incandescent light bulb, and direct and average daylight, as defined by the CIE.

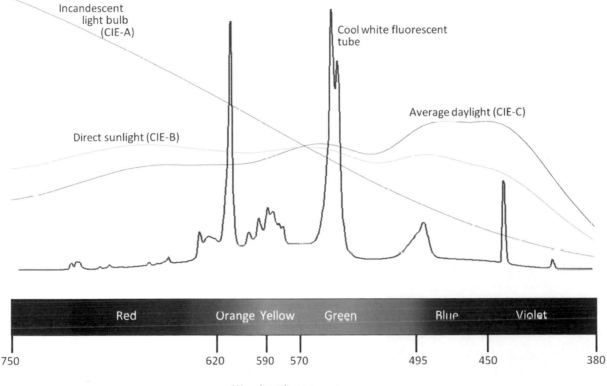

Figure 64: Spectra of some light sources (*fluorescent spectrum adapted from Padleckas, 2005*)

You can see that the fluorescent tube has a narrow burst of orange light and green light, together with a narrow, lower-energy blue-violet component. There is almost no red, yellow or mid-blue. Conversely, both daylight and incandescent light are smooth and continuous, with energy over the whole visible spectrum.

There are some other very interesting color illusions that arise from the preprocessing and interaction that take place between adjacent receptors in the eye. We will explore these further when we discuss seeing patterns.

It is worth mentioning that the eye preprocesses the color information to a much lower resolution than brightness.

This limitation of the eye is taken advantage of in storing and transmitting pictures - still or moving. The brightness or "black and white" content is separated from the color content and then the color content is blurred to minimize the amount of data to store or transmit. This is true for both analog and digital television, and digital photography, including the various JPG and MPG formats.

Finally, the eye has no blue receptors in its very center - the foveola mentioned earlier - and there are no black-and-white rod receptors either.

Here's a simple demonstration of the lack of blue receptors in the center. Prop the book up and look at the 3 circles with their colored stars.

Figure 65: No blue receptors in our central vision

Now slowly move away from the picture. The blue star will quickly disappear into the black, while the red and green stars remain easily visible.

Color Constancy

A sheet of white paper appears white both in bright sunlight, and when we see it under dim white light. Similarly, that same sheet of white paper appears white both in bright sunlight, and when seen under dim incandescent light.

It does not appear a dull orange, even though, as we showed earlier, dim incandescent light has far more energy in the yellow-orange part of the visible spectrum. It is the relative color of items in a scene that matters - so-called color constancy.

In the real world, the average color of a scene across all items commonly approximates a neutral gray.

This "integration to gray" is one way a modern camera attempts to auto-correct for different illumination color and still provide a reasonably balanced-looking picture - so-called automatic white balance.

(a) A camera's automatic white balance failure to perceptually reproduce colors in a scene under incandescent light.

(b) the same scene photographed under daylight illumination and displaying correct colors

Figure 66: Color constancy failure in a camera

Of course this is not a perfect solution since there can be huge variations in brightness and color of the illumination, and scene content may have a dominant color - e.g. when photographing flowers close up.

In practice, we have all experienced the disappointment of a camera photo with a strong orange cast caused by incandescent lighting, while our eyes gave us no hint of any problem.

The human eye isn't so easily fooled and normally interprets the colors in the scene satisfactorily under either form of illumination.

Color Contrast

Similar to seeing brightness, the eye must cope with all the varied intensities and wavelengths of light from the scene.

Then it must make judgments on the content, judgements that are heavily influenced by local and overall context.

Again, such judgments can be erroneous and so lead to illusions. To illustrate the point, the next figure shows a colored version of our gray flock of birds illusion.

You can see that the bird color appears to change from pale blue on the left, to pale gold on the right, shifting quite abruptly around the transition from sun to sky.

Yet the birds are all an identical neutral gray.

Figure 67: The birds are all the same neutral gray in color

Now let's return to our black-and-white checkerboard illusion.

This time we will overlay a colored circle on each of the two target gray squares.

We still perceive the two gray squares as having different brightness, even though we know they are the same.

In addition, we now perceive the two colored circles as being somewhat different in color and brightness.

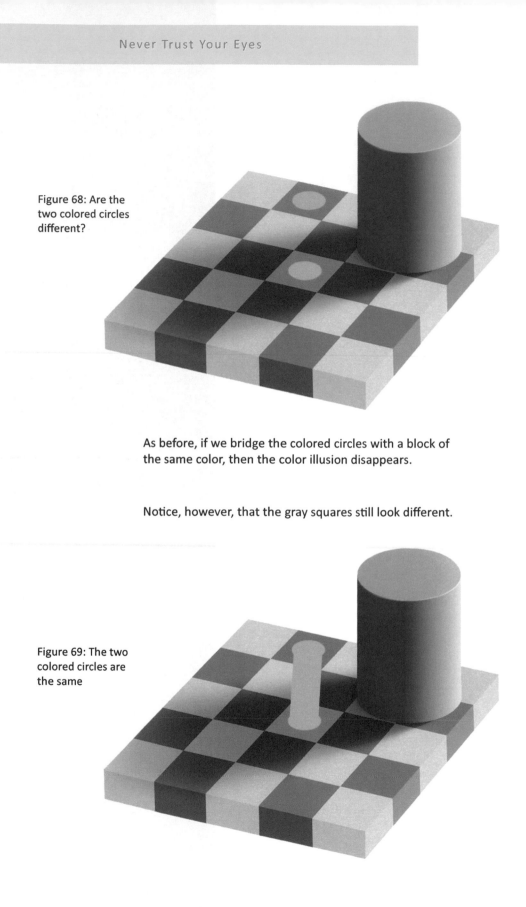

Figure 68: Are the two colored circles different?

As before, if we bridge the colored circles with a block of the same color, then the color illusion disappears.

Notice, however, that the gray squares still look different.

Figure 69: The two colored circles are the same

Seeing at Night

As already noted, the eye has two types of receptors - cones and rods. The cones are responsible for our daylight, color vision and are concentrated in a small area in the center of the eye.

The rods are responsible for vision in dim light, and exist over the whole surface of the eye's image sensor, except for the very small central part where we have best daylight, color vision.

This absence of rods in our central vision explains why at night we feel the need to avert our gaze slightly to best see. For example, if you look directly at the star constellation Pleiades, or the Seven Sisters, you will find it difficult to see seven stars. However, if you avert your gaze slightly, you will see more than seven.

Figure 70: The Pleiades or Seven Sisters
(NASA/ESA/AURA/Caltech)

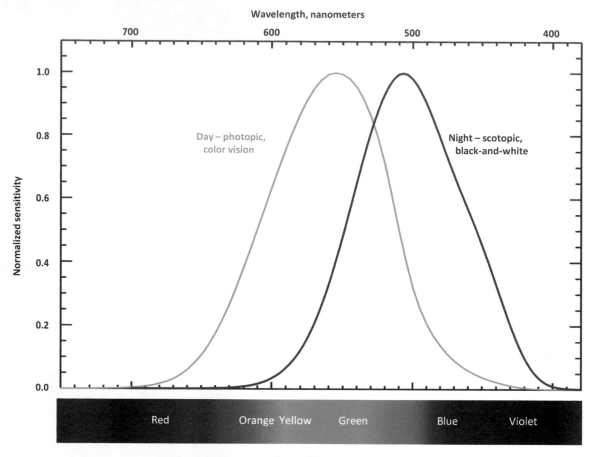

Figure 71: Our color shift from day to night vision
(from the CIE definitions)

The switch from using our cone-based daylight vision to rod-based nighttime vision does not happen quickly. The chemical processes that operate the rods in the eye's image sensor take about 40 minutes to fully adapt to low light levels. This explains why we are almost completely blind when we enter a dark cinema during daylight hours and collide with the rows of seats.

Our rod-based vision is only black-and-white, no color is discernible. Additionally, the chemical properties of the rods make them most sensitive to light that is towards the bluer part of the spectrum. This color shift, called the Purkinje shift, explains why some objects can take on a rather different appearance at night.

This effect is most noticeable when viewing, say, red and blue flowers. In daylight, the red may stand out as well as the blue flower against the green leaf background, but as night falls, the red flower will lose color and darken well below the green leaves, while the blue flower will appear to brighten against the green leafy background.

(a) daytime appearance   (b) nighttime appearance

(a) daytime appearance   (b) nighttime appearance

Figure 72: Differently colored flowers viewed in daylight and at night

As an aside, since our rod-based vision is insensitive to red light we can use bright red lights without reducing our ability to see at night. This technique is sometimes used in car and airplane dashboards where the display is bright red and is seen by the red (L) cones, while our sensitive rods continue to provide good nighttime vision.

Seeing Patterns and Movement

The eye is continually searching the scene and tracking items of potential interest to make sense of what it is seeing.

At left is a well known ambiguous drawing which, by simple suggestion, we can help your brain switch from one view to the other view, or even see both views at the same time.

Is it a swan, a squirrel, or both?

Context, experience and suggestion all play their part.

Figure 73: Simultaneous ambiguous illusion of a swan facing left or squirrel right *(G H Fischer, 1968)*

In this next classic illusion, is this an old woman or a young girl?

Here it is hard to see both simultaneously.

Rather the impression switches back and forth between the two possibilities.

Figure 74: Alternating ambiguous illusion of young or old woman *(W. E. Hill, 1915 after anonymous German postcard 1888)*

Figure 75: Better to mistakenly identify a tiger, when
there isn't one, than not identify it and get eaten
(Tiger in the Valley © D L Rust)

A stable, static scene poses little danger and so is of little interest, but any change must be quickly detected, and its form swiftly identified as **FRIEND, FOE, or FOOD**.

To ensure our survival in the real world, the eye has evolved real-time motion and pattern recognition capabilities.

Real-time of course because it's of no use to our survival if we detect that a tiger was leaping at us minutes after the fact.

So, while pattern and movement recognition are very complex processes, the eye (and brain) must continually, in real time, extract scene information and make best guesses as to what is happening.

These guesses can go wrong, and we call these errors illusions.

But it is better for our survival if we err on the safe side and guess that it is a tiger, even if occasionally we're wrong and there is no clear and present danger.

So the eye is continually assessing the scene and labeling what it sees - then updating and correcting, as more information becomes available. Was it a tiger in the distance, or was it a black and yellow wasp nearby?

* * *

In certain circumstances we will mistakenly perceive movement in stationary patterns, and patterns where there is only movement.

The next two sections explore our ability to see pattern and movement, realizing that they are inextricably linked.

Seeing Patterns

A static pattern in its most basic form is just a change in the brightness and/or color as we move from one area of the scene to another.

So the eye first identifies and extracts edges - the transitions from one brightness or color to another.

Detecting an edge and then interpreting that edge are key first processes. Here is a classic illusion to illustrate the point.

Look at the picture and what do you see? Two somewhat different gray squares?

Figure 76: Two different gray squares?
(Cornsweet illusion, Fibonacci, 2007)

Now place your finger over the line dividing the two squares. Now what do you see? The two squares are actually the same gray.

The eye has been fooled by providing it with a transition that it could have extracted had there really been two different gray squares. The eye treats the edge as critical information that overrides the fact that the brightness of each square is in reality the same.

101

As we showed earlier, just knowing the brightness or color of an item is not very useful in the real world with its huge variations in light and color. The eye adjusts to global differences, and then seeks local differences in brightness and/or color.

Ultimately, it is the local contrast in a scene that helps us make sense of each item's shape and texture, and enables us to judge them appropriately.

Given the generally unstructured nature of the real world, it is not surprising that the eye has difficulty with very structured patterns.

Usually our eye and brain have no need to cope with such structure.

Here is a classic example of how the eye misinterprets a very structured pattern.

From top to bottom, we are simply offsetting the middle row little by little.

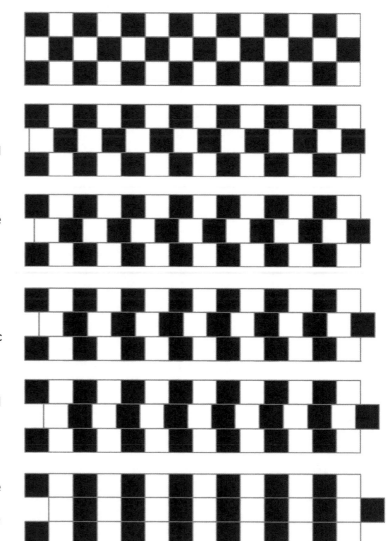

Figure 77: Skewed or parallel lines?

Notice how skewed we perceive the horizontal lines.

Take a ruler to them and you can verify they in fact remain quite straight.

However a color version shows very little effect, provided the colors are not strongly contrasted, as below.

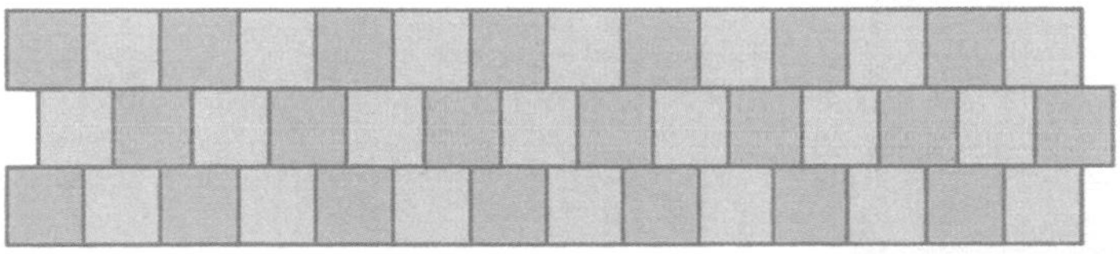

Figure 78: Mostly parallel lines

Clearly the eye dislikes strong patterns, but conversely, the eye abhors emptiness and will continually seek to make sense of the scene. Here the eye attempts to fill the gap with a non-existent star pattern.

Figure 79: There is no star
(inspired by the well known Kanizsa triangle illusion)

103

Seeing Movement

As mentioned earlier, in certain circumstances we will perceive movement in stationary patterns, and patterns where there is only movement.

A simple example of the ambiguity of pattern and movement is that of two alternately flashing lights. As the flash speed changes, the eye will perceive not two separate flashing spots but just one spot, leaping from left to right and back again.

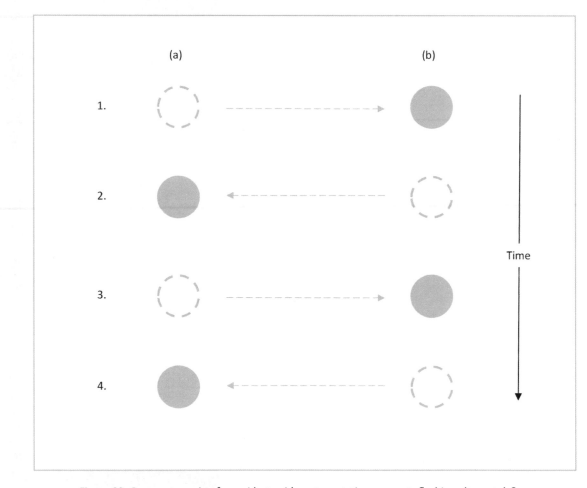

Figure 80: One spot moving from side-to-side or two stationary spots flashing alternately?

Of course, it is hard in a static book to illustrate movement, but the point is that brightness and color over the scene and their changes over time and space provide the raw information we need to be able to detect movement and identify patterns in the scene.

Once we have that raw information, we need to process it quickly and effectively. The eye does this through a massively parallel set of pre-processing neurons in the eye itself - and sending the brain a number of different scene extractions[5] for even more sophisticated processing.

The eye can mistakenly find movement in static patterns, and there are many illusions that show this well. A good example is illustrated below. However, in the real world we rarely encounter such heavily structured patterns.

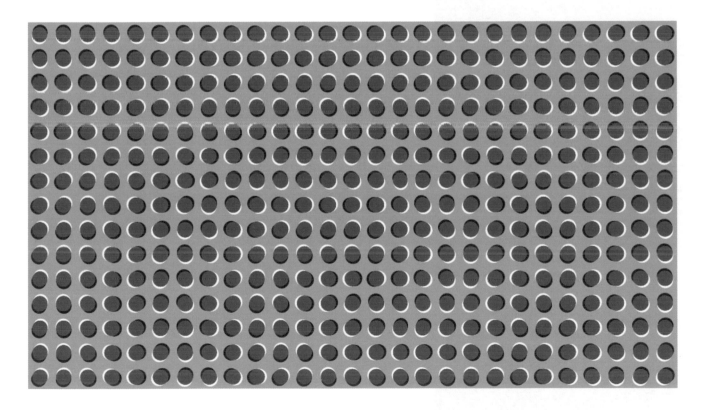

Figure 81: Illusory movement in a static pattern

Television and Cinema

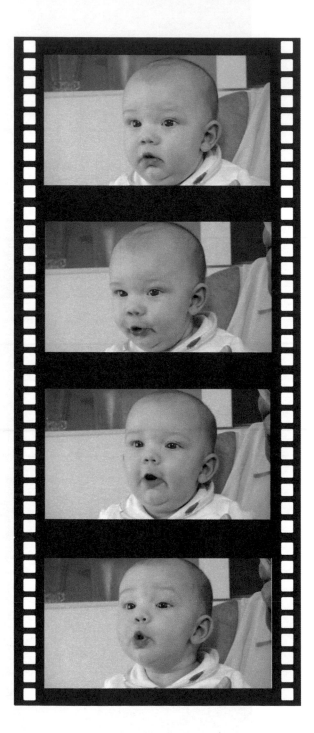

The earlier illusion of the moving dot - or was it two dots flashing - may give you a clue as to how the illusion of movement is created in television or cinema.

The video or movie camera takes a series of still images, as illustrated in the figure. Each image is simply a snapshot of the brightness and color in the scene at that instant.

By showing the still images in sequence, directly on a screen or via a projector, the eye is presented with the information it needs - the changes in brightness and color from one moment in time to another.

Around 25 to 30 still images each second are generally enough to give the illusion of movement.

Typically each image is repeated 2 or 3 times to help the eye smooth out or integrate the brightness changes, and provide a steadier appearance, especially in bright uniform static scenes.

Figure 82: How movies work

Rotating Wheels

You have probably seen wagon wheels on TV appearing to go slowly backwards, when clearly the wagon itself was heading forwards. The early pioneers were definitely rolling fast and furious Westward, and not slowly returning to the East.

Like wagon wheels, modern-day car wheels are often highly structured and show the same reverse rotation illusion - because this is an illusion, and you DO NOT see it in the real world.

It is purely an artifact due to the way in which TV and cinema work by taking a series of stills or snapshots, and not truly showing continuous movement.

Here's what happens.

 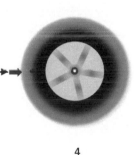

| 1 | 2 | 3 | 4 |

Figure 83: Wheel rotating - but forwards or backwards?

The figure shows a wheel that has rotated and we have two snapshots - one at time 1 and one at time 4. The red dot and arrow are simply there to help show where the wheel is at any one time.

So, did the wheel roll forwards or backwards?

If we assume it rolled forward then we fill in the blanks at time 2 and time 3 as below.

 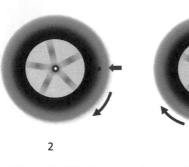 

1   2   3   4

Figure 84: Wheel rotating forwards by 3/4 turn

But did the wheel roll forward?

It could just have easily rolled backward, only more slowly, as shown below.

 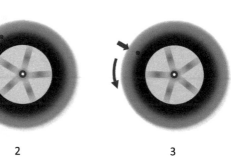 

1   2   3   4

Figure 85: Wheel rotating backwards by 1/4 turn

The wheel has still finished in the same position - we just made a different assumption about the direction in which it was turning and at what speed it was rotating.

In this case the wheel appears to be going backward at about one third the speed of its apparent forward motion.

In the same way, this is how our eye works when presented with two snapshots from a video sequence. It takes the change in pattern from one snapshot to the next and first recognizes the change as motion.

Then it has to decide whether that motion is fast forward or slow backward. Sometimes it seems to make more sense interpreting the wheel going backward, despite other clues in the scene strongly to the contrary.

Curiously, there is one known instance in the "real" man-made world where we can see this illusion.

Street lights such as high-pressure sodium lamps produce artificial lighting that is pulsating in nature - a sharp burst of light followed by a period of darkness, and so on.

Essentially the lamp extinguishes at each zero-current crossing in the AC cycle, so the light goes on and off either 120 times per second (for 60Hz countries) or 100 times per second (for 50Hz countries).

So this form of lighting illuminates the scene in bursts, providing the eye with snapshots, much like video sequences. A moving car's wheels, viewed under such light, can be perceived as rotating forward or illusorily backward.

This of course does not happen in daylight which is a continuous, steady stream of light, unlike the light from these modern-day street lamps.

Seeing Depth

With our forward-facing eyes giving us overlapping views of the same scene, we have the great advantage over many other animals that we see in three dimensions. This stereo vision, from the Greek word *stereós* meaning solid, provides us with a superb sense of depth perception and gives us the close dexterity needed to best use tools and manipulate objects.

It is important to realize that perceiving depth does NOT require stereo vision, since there are many clues to depth perception that require only one eye, so-called monocular vision.

But with two eyes, giving us overlapping views, we have greatly enhanced near precision and, as anyone who has recently been to the cinema and experienced a movie in 3D, the strength of our stereo vision can be spectacular.

Seeing with One Eye

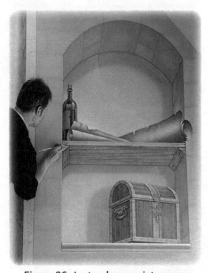

Just as with seeing patterns and movement, context and experience play important roles in our perception of depth.

Here's a simple example.

Is this just a drawn picture or a real scene?

In this case we're not easily fooled. We have a clue that the original scene was just a picture since the man appears to be drawing it.

Figure 86: Just a drawn picture or a real scene? *(© Funatico.com, 2008)*

On the other hand, this illusory picture, combining a real street scene with a stunning chalk drawing on the pavement, is a great example of perceiving depth while only needing one eye.

Figure 87: Pavement art LavaBurst
(© 2008 Edgar Mueller)

In this case, it is much harder to decide which parts are real and which artificial.

So first, let's explore the numerous depth clues we have with a single eye and how many of these monocular depth clues are used in the pavement art.

Perspective

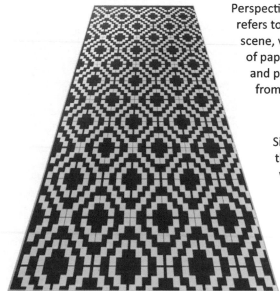

Figure 88: Perspective

Perspective is a very strong depth clue. It refers to the fact that similar items in a 3D scene, viewed on a flat surface like a sheet of paper, exhibit changes in relative size and position, depending on their distance from the viewer.

Similar items that are larger give us the sensation of their being nearer, while smaller ones appear farther away. Here we show the pattern in a carpet appearing to get smaller.

A good example of perspective in the pavement art is the larger size of the row houses nearby, while farther ones become smaller and smaller, and closer and closer together - although, of course, we know in reality that they are all of a similar size and distance apart.

We can envision notional lines running along the houses - the sidewalks - that in reality are equally apart, but in the picture these lines would converge in the distance.

Even if there are unknown items in the scene, if they are of different sizes, then we will generally make the assumption they are of the same size but at different distances.

Occlusion

Occlusion simply refers to the fact that a nearer item can partially or completely obscure a more distant one.

In the pavement art, there are many examples of occlusion - real and artificial. These include the people standing over the gorge and obscuring some of the water below, and pedestrians on the right partially obscuring the houses.

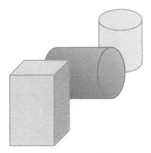

Figure 89: Occlusion

Aerial Perspective or Haze

As items get farther away, they tend to become softer and slightly blue tinged. The air between us as observers and the distant item plays a major part in this, along with our diminishing visual acuity over distance.

In the picture, there is softening and lightening of, for example, the trees along the sidewalk, reinforcing the sense of depth and reality.

Figure 90: Haze

Light, Shade, Texture and Gradient

Even completely smooth, uniform items show changes in their appearance due to the play of light and shadow.

If the item has texture or variations in shape, these aspects will cause the light and shadow to provide good depth clues.

Figure 91: Light, shade and texture

In the pavement art picture, you will notice that the real scene's left side is sunny while the right is in shade. For the pavement art itself, there is strong texture, combined with light and shade in the gorge's cliffs and in the rippling water, so reinforcing our sense of depth and reality.

Motion Parallax

As *we* move, even stationary items in a *real* scene appear to move.

If we move to our right, nearer items appear to move more, but in the opposite direction, while distant ones like the moon or stars seem to keep pace with us.

In the pavement art, since it is a printed flat image, no movement on our part is going to give us this clue. The fact that we *don't* see this motion parallax suggests to us this is not a real scene.

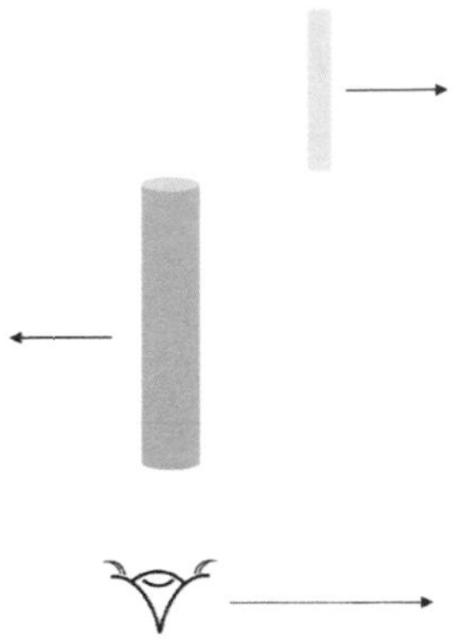

Figure 92: Motion parallax

Motion Depth

This is similar to motion parallax, but refers to *items* in the scene moving, as opposed to us as the observer moving.

In a real scene, we can readily perceive side-to-side movement of items. However, if an item is moving directly towards us, only its relative size increases. The eye has to quickly interpret this size change as meaning the item is coming towards us - and take avoiding action if necessary. Think about how quickly you can assess a vehicle coming towards you.

Again, since the pavement art is only a static image on paper, no movement is going to occur - the fact that we don't see any motion suggests this is not a real scene.

Accommodation

This refers to the amount of effort being expended by the muscle around the eye's lens. As explained earlier, the lens becomes more rounded to focus on nearer items.

Again, since the pavement art is only an image on paper, there can be no difference in the accommodation effort needed to focus on near or far items, and so this depth clue is not available.

* * *

In addition to the above clues, our *prior* experience plays a very strong role in perceiving depth.

This next drawing illustrates the point. Figure (a) looks like what it is - a collection of flat lines forming a simple two-dimensional, cut diamond pattern.

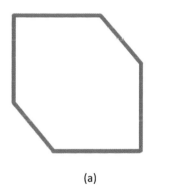
(a)

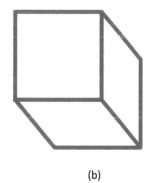
(b)

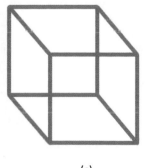
(c)

Figure 93: Flat or three-dimensional figures?

Yet if we add just a few lines, as in (b), we perceive a three-dimensional cube - even though we know it is still just a collection of flat lines on a flat piece of paper.

Add a few more lines, as in (c), and we get a wire frame outline of the so-called Necker cube that is a wonderfully ambiguous three-dimensional object. Which is the front and which is the back face of the cube?

Here is a splendid version of the Necker cube constructed from votive candles on a floor.

Figure 94: Necker cube of candles viewed from one end *(Amazing Fire Illusion, © Brusspup, 2010)*

When seen from exactly the right viewpoint, the set of candles on the flat two-dimensional floor suddenly bursts into a fully three-dimensional, illuminated cube.

Figure 95: Necker cube of candles viewed from the precise perspective *(Amazing Fire Illusion, © Brusspup, 2010)*

I recommend the YouTube video of this, as the camera slowly moves around the room to the right perspective.

Because depth perception is a complex exercise, our perception of depth, even in real world scenes, can be easily compromised.

Here is a superb example where the three vehicles look very different in size, yet if you measure them with a ruler you will find them to be identical.

Figure 96: Real world size illusion - all the vehicles are the same size
(earthguide.blogspot.com, 2009)

The eye is making a huge assumption, based on the perspective clues from the road and path, along with partial occlusion of the vehicles, to falsely estimate that the vehicles are quite different in size.

Of course, this illusion is only be temporary in reality. As we move around in the scene we soon realize that the three vehicles are the same.

Let's finish this chapter with another superb example of
a *trompe l'oeil* (French for trick of the eye). Unlike the
pavement art we showed earlier, which was drawn on the
ground, this artist paints murals.

Figure 97: Trompe l'oeil mural - can you see what is real and what is painted?
(Montpellier, France, madart.fr, 2007)

Again, the art is integrated with reality. This is a real
building, a few of the windows are real, and some of the
people, tables and chairs of the pavement cafe are real.

Seeing with Both Eyes

We just described the many clues to depth perception that exist using a solitary eye.

None of these clues compares with the stunning beauty of true stereo vision - the ability of our two eyes to take two disparate views of the same scene and fuse them into a beautiful, solid representation of the real world.

When we use both our eyes, termed binocular vision, there are two additional clues to depth. These are described below, and illustrated in the figure.

Stereopsis

This is the ability of our eyes to provide two simultaneous, but somewhat different, views of the same scene, together with the brain's ability to fuse them into a single, solid image.

In the figure, we have represented the two views of a cube in two different colors and coded the direction of gaze for each eye accordingly.

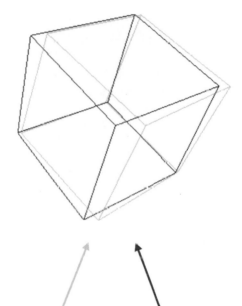

Left eye   Right eye

Figure 98: Seeing with two eyes

119

Convergence

This is simply the amount by which each eye departs from the parallel to focus on a near object.

In the previous figure, you can easily observe that each eye is turned in toward the object being viewed.

* * *

From the above, clearly stereo vision requires significant overlap of each eye's field of view.

In general, predators require good stereo vision to precisely locate their source of food and so have forward facing eyes with overlapping fields of view. However, animals of prey, like the rabbit, have a greater need to be able to detect a predator anywhere around them, so have side-facing eyes, providing a tremendous field of view, being able to see almost 360 degrees, but with no overlapping field of view to provide stereo vision.

Conversely, a rabbit does not need good close vision like a predator to find its food source, since it eats grass which is readily available, literally under foot.

3D Image Examples

In a book it is of course difficult to illustrate 3D but I have included an example anaglyph that can be viewed using red and cyan filter glasses available from many companies at very low cost.

Much like the previous cube figure demonstrated, an anaglyph comprises two overlapping views of the same scene, just like our eyes would see. One image is then colored cyan, and the other red.

Using the colored 3D glasses, each eye essentially sees only one of the overlapped images.

Figure 99: sample anaglyph glasses for the picture below

Figure 100: Anaglyph of Saguaro National Park at dusk
(U.S. Geological Survey)

121

The two separate images are then fused in the brain to give us a sensation of solidity and depth.

This red-cyan anaglyph approach renders color poorly. However, movies are now being released in 3D, using a similar technique to the above, but replacing the colored filters with differently polarized filters of a neutral gray that allow full color rendition.

Some television systems are also starting to provide affordable, high-quality 3D viewing at home. Again, these use the same basic approach of presenting a separate image to each eye, and still requiring the user to wear some form of glasses.

Unlike 3D movies, current 3D-TV systems actively switch each eye's lens on and off alternately, in synchronization with the left and right images presented on the television screen. At the time of writing, some manufacturers are also planning to offer systems using passive, polarized glasses.

Finally, digital still and movie cameras offering high quality 3D, using twin lenses and image sensors, are also now appearing on the market.

* * *

In the meantime, here is a 3D picture that you *can* see on the printed page, *without* using glasses.

Turn the book clockwise to view the picture in landscape format. Hold the book at arm's length, and relax your eyes as though looking *through* the picture at a distant scene.

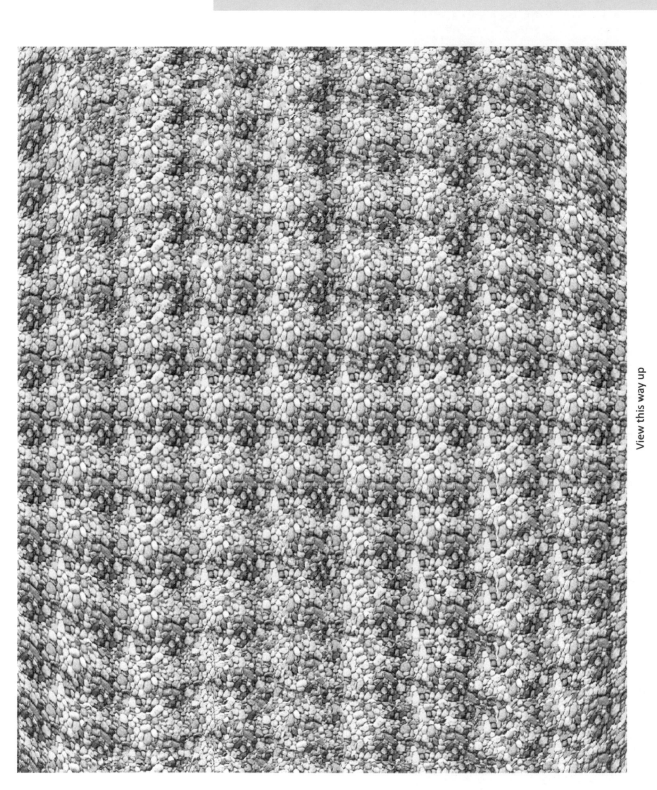

View this way up

Figure 101: Autostereogram of a flock of birds (turn the book to view)

After a few moments the random pattern should suddenly snap into a 3D scene of a flock of birds, some nearer, some farther away.

If you find this difficult at first, or if the image appears broken up, simply close your eyes and then open them very slowly so that you start seeing the picture initially as a dim blur. The 3D scene should now appear.

These types of 3D images are called autostereograms. They are more difficult to see than 3D images using glasses, of course, so don't be concerned if you cannot see anything at first.

Also, for various reasons, not everyone has stereo vision, so you may not be able to see true depth in 3D movies, or any of the 3D pictures presented here.

For those who need a clue, below is a black-and-white figure to help you identify what you will see in the autostereogram.

Figure 102: What you should be able to see in the autostereogram

Conclusion

So ultimately, is the eye just like a high-quality digital camera?

I hope you can now see that the eye and brain together could only be matched by a digital, full color, 3D, super high resolution movie camera that is continually and automatically sampling and tracking areas of interest in the scene, while simultaneously auto-focusing, auto-exposing, auto-processing and interpreting the images received.

That we are very far from this capability is demonstrated by the fact that robot vision is still very rudimentary.

Unlike humans, robots still cannot freely move around in space and time without continually coming to grief.

Need I say more?

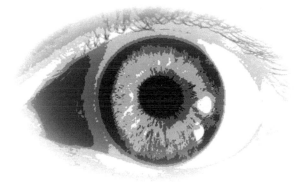

Figures and Credits

Figures and credits are listed below, and are all courtesy of the author, unless otherwise indicated. Every effort has been made to obtain copyright release but if any inaccuracy or omission is found, please contact the publisher for correction.

Figure 1: A digital camera's uniformly detailed, colorful image; Figure 2: Illustration of what the eye sees instantaneously ; Figure 3: Sketch of an early camera obscura *(Fizyka,1910)*; Figure 4: Illustration of a full-size camera obscura with a panoramic rotating mirror *(adapted from A. Rees, Cyclopoedia Universal Dictionary of Arts and Sciences 1778, © National Maritime Museum, Greenwich)*; Figure 6: The splendid view of the outside from the inside of the Edinburgh camera obscura *(© Camera Obscura and World of Illusions, Edinburgh)*; Figure 5: The wonderfully ornate architecture of the full size camera obscura in Edinburgh, Scotland *(© Camera Obscura and World of Illusions, Edinburgh)*; Figure 7: Human eye front view; Sclera; Figure 8: The eye as a basic camera; Pupil; Iris; Cornea; (transparent opening); Iris; (aperture); Lens; Figure 9: Basics of the human eye; Sclera; (light-proof enclosure); Retina; (image sensor); Optic Nerve; (multi-wire connection to the brain for image processing and storage); Figure 10: Atria Belladonna *(adapted from Koehler's Medicinal-Plants 1887)*; Figure 11: The same image but with small pupils in the upper picture, and large pupils in the lower picture - is there is a difference in attractiveness?; Figure 12: A modern digital compact camera; Figure 13: Image creation in a modern-day digital camera; Aperture; Figure 14: The basics of a modern-day digital camera; Lens; Light-tight enclosure; Digital memory and processor; Shutter; Multi-wire connection; Image sensor; Figure 15: Illustration for the eye's blind spot; Blind Spot; Figure 16: Location of the eye's blind spot; Figure 17: The true size and shape of the eye's blind spot; Figure 18: Blind Spot - demonstration of the brain filling in our gap in vision; Image from each eye with its blind spot; Reconstructed image; Original scene; Left Eye; Right Eye; Figure 19: The blind spot in the right eye is on the opposite side from that of the left eye; Figure 20: Expanded view of the eye's light sensitive surface (Retina); Direction of light from the scene; Retinal blood vessels; Optic nerve fibers; Pre-; processing cells; Light ; sensitive cells; Cones (for daylight ; color vision); Rods (for nighttime vision); Light from the scene; Color filters; Figure 21: Basics of a camera's image sensor; Image sensor; Figure 22: The camera's uniformly detailed, colorful image of a real scene; Figure 23: Simplified illustration of the eye's image sensor; Cone receptors for color vision; Rod receptors for black and white nighttime vision; Figure 24: The eye can only see detail in the tiny area called the Fovea; Fovea; Foveola (cone-based color receptors only); Macula; 0; 0.5; millimeters; Fovea; 1.0; Figure 25: Magnified view of the eye's "super sensor" area *(adapted from Padgham & Saunders, 1975)*; 1.5; Figure 26: We only see detail over this small area; Figure 27: Simulation of what the right eye sees instantaneously ; Figure 28: Our eyes ; cannot read the letters; Figure 29: Playing card demonstration; Figure 30: Overall blur due to low light level and camera shake; Figure 31: Partial blur due to some items moving during exposure; Figure 32: Blur of surround due to following a moving item *(© Matthias Trischler, 2009)*; Figure 33: Near blur from a traveling camera; Figure 34: Blur from limited depth of field; Figure 35: Image fading - the Troxler effect; Figure 36: Afterimage demonstration of receptor saturation; Figure 37: The eye absorbs detail in a small area ; then jumps to the next area and so on; Figure 38: Can you see your own eyes move?; Figure 39: But can you see their eyes move?; Figure 40: How we read; Figure 41: Our instantaneous reading ability *(Hans-Werner, 2008)*; Figure 43: Camera set for near focus - the lens moves forward towards the object; Figure 42: Camera set for distant focus; - the lens moves back away from the object; Figure 44: Eye lens adjusted for distant focus; - the lens is not so curved; Figure 45: Eye lens adjusted for near focus - the lens is much more curved; Figure 46: Near sight (Myopia) and its correction; Figure 47: Far sight (Hyperopia) and its correction; Figure 48: A camera's aperture set open (a) wide and (b) small *(adapted from Mohylek, 2006)*; Figure 49: The eye's aperture or iris open (a) wide and (b) small *(adapted from Nicko Va, 2009)*; Fovea; Figure 50: View into the eye showing the back inside surface and the retinal blood vessels *(adapted from Ignis, 2007)*; Blind spot; Figure 51: Example of the red eye effect from a camera flash *(PlePle2000, 2006)*; Figure 52: Flying birds are all the same gray; Figure 53: Squares A and B are the same *(Adelson, 1995)*; Figure 54: Do the squares A and B still appear different?; Figure 55: Squares A and B really are the same; Figure 56: Uniform gray bars seem shaded (Mach bands); Figure 57: The range of visible light; Figure 58: How our three color receptors respond to light *(adapted from Stockman and Sharpe, 2000)*; Figure 59: How our three color receptors respond to colors in a scene; Figure 60: Additive and subtractive color mixing; Figure 61: True yellow and a red-green mix simulation; Figure 62: Horizontal traffic lights; Figure 63: Example Ishihara color test chart; Figure 64: Spectra of some light sources *(fluorescent spectrum adapted from Padleckas, 2005)*; Figure 65: No blue receptors in our central vision; Figure 66: Color constancy failure in a camera; Figure 67: The birds are all the same neutral gray in color; Figure 68: Are the two colored circles different?; Figure 69: The two colored circles are the same; Figure 70: The Pleiades or Seven Sisters *(NASA/ESA/AURA/Caltech)*; Figure 71: Our color shift from day to night vision *(from the CIE definitions)*; Figure 72: Differently colored flowers viewed in daylight and at night; Figure 73: Simultaneous ambiguous illusion of a swan facing left or squirrel right *(G H Fischer, 1968)*; Figure 74: Alternating ambiguous illusion of young or old woman *(W. E. Hill, 1915 after anonymous German postcard 1888)*; Figure 75: Better to mistakenly identify a tiger, when; there isn't one, than not identify it and get eaten *(Tiger in the Valley © D L Rust)*; Figure 76: Two different gray squares? *(Cornsweet illusion, Fibonacci, 2007)*; Figure 77: Skewed or parallel lines?; Figure 79: There is no star; (inspired by the well known Kanizsa triangle illusion); Figure 78: Mostly parallel lines; Figure 80: One spot moving from side-side or two stationary spots flashing alternately?; Figure 81: Illusory movement in a static pattern; Figure 82: How movies work; Figure 83: Wheel rotating - but forwards or backwards?; Figure 85: Wheel rotating backwards 1/4 turn; Figure 84: Wheel rotating forwards by 3/4 turn; Figure 86: Just a drawn picture or a ; real scene? *(© Funatico.com, 2008)*; Figure 87: Pavement art LavaBurst *(© 2008 Edgar Mueller)*; Figure 88: Perspective ; Figure 89: Occlusion ; Figure 90: Haze ; Figure 91: Light, shade and texture ; Figure 92: Motion parallax; Figure 93: Flat or three-dimensional figures?; Figure 95: Necker cube of candles viewed from the precise perspective *(Amazing Fire Illusion, © Brusspup, 2010)*; Figure 94: Necker cube of candles viewed from one end *(Amazing Fire Illusion, © Brusspup, 2010)*; Figure 96: Real world size illusion - all the vehicles are the same size *(earthguide.blogspot.com, 2009)*; Figure 97: Trompe l'oeil mural - can you see what is real and what is painted? *(Montpellier, France, madart.fr, 2007)*; Figure 98: Seeing with two eyes; Figure 100: Anaglyph of Saguaro National Park at dusk *(U.S. Geological Survey)*; Figure 99: sample anaglyph glasses for the picture below; Figure 101: Autostereogram of a flock of birds; Figure 102: What you should be able to see in the autostereogram

Text References

1. Megapixel rating of the eye: http://www.clarkvision.com/imagedetail/eye-resolution.html

2. Cephalopod lens focus: *Brain and Behavior in Cephalopods*, M J Wells, Stanford University Press, 1962

3. Color shades: *Color in Business, Science and Industry,* Judd, Deane & Wyszecki, Wiley, 1975

4. Color Blindness: *More Prevalent Among Males*, Howard Hughes Medical Institute, http://www.hhmi.org/senses/b130.html, retrieved 2010-11-22

5. Scene extraction: *The Movies in Our Eyes,* Werblin and Roska, Scientific American, 296, 4, April 2007

Further Reading

1. *The Reproduction of Color*, R W G Hunt, Wiley, 2004

2. *The Perception of Light and Colour,* C A Padgham and J E Saunders, Bell, 1975

3. *Seeing: Illusion, Brain and Mind*, John P Frisby, Oxford University Press, 1979

4. *Eye, Brain and Vision,* David H Hubel, Scientific American Library, 1988

5. *The Artful Eye,* edited Gregory, Harris, Heard and Rose, Oxford University Press, 1995

6. *Art and Illusion*, E L Gombrich, Phaidon, 1993

7. *A Natural History of Seeing*, S Ings, Norton, 2008

8. *Eye and Brain* by R L Gregory, Oxford University Press, 1998

9. *Vision and Art - The Biology of Seeing,* Margaret Livingstone, Abrams, 2002

10. *The Science of Perception - 169 Best Illusions*, Stephen L. Macknik and Susana Martinez-Conde, Scientific American MIND, June 2010

About the Author

During his research into Visual Telecommunications at British Telecom, Trevor White became fascinated with how our eyes work, and the illusions that arise from the practical constraints that Nature selected for the eye.

In this book he explains the basics of the eye and compares and contrasts it with the modern-day digital camera that can take both stills and movies.

Using amazing visual illusions, he demonstrates the eye's huge limitations compared with a camera - but then he shows how Nature splendidly overcomes the limitations to give us the ultimate illusion of near-perfect sight, that is well-matched to our survival in the real, living world.

Trevor White holds a Master's degree in Visual Science, a Master's degree in Business Administration and a Bachelor's degree in Electronics. Previous publications in professional journals include *Nature* and *Proceedings of the IEE*.